Control Surface: Identity Governance for AI Agents

Practical Frameworks for Securing Agent Identities, Dynamic Permissions, and Auditable Control in Enterprise AI

JC Louis-Charles

Cybersoft Publishing LLC

Fort Washington, MD 20744

Drclaude.net

First Edition March 2026

Contents

Introduction 1
1 Why Identity Fails First 7
1.1 Scenario: The Pilot That Spiraled 7
1.2 Why It Matters 8
1.3 Framework: The Four Identity Fracture Points 12
1.4 Example: A Fed. Agency's AI Onboarding Misstep.19
1.5 Manager's Checklist 22
1.6 Takeaway 25
2 Human, Machine, and Agent Identities 28
2.1 Scenario: The Service Account Assumption 28
2.2 Why It Matters 29
2.3 Framework: Defining Each Identity Class 33
2.4 Example: Healthcare Plat. Identity Reclassification.41
2.5 Manager's Checklist 44
2.6 Takeaway 46
3 Identity Architecture for Agentic AI Systems 48
3.1 Scenario: Arch. Review That Revealed Everything .48
3.2 Why It Matters 49
3.3 Framework: Core Architectural Components 53
3.4 Example: Evaluating an Arch. Proposal 62
3.5 Manager's Checklist 65
3.6 Takeaway 68
4 Dynamic Permissions & Deleg. Autonomy 70
4.1 Scenario: The Agent That Promoted Itself 70

4.2 Why It Matters 71
4.3 Framework: Managing Dynamic Permissions 75
4.4 Example: A Fin. Serv. Firm's Permission Overhaul . 82
4.5 Manager's Checklist 85
4.6 Takeaway 87
5 Identity Lifecycle for AI Agents 89
5.1 Scenario: The Zombie Agent 89
5.2 Why It Matters 90
5.3 Framework: The Seven Lifecycle Stages 94
5.4 Example: Scaling Lifecycle Governance at Cloud 105
5.5 Manager's Checklist 107
5.6 Takeaway 109
6 Identity Risks Unique to AI & Multi-Agent Sys. ... 112
6.1 Scenario: The Silent Impersonation 112
6.2 Why It Matters 113
6.3 Framework: The Six Identity Risk Categories 118
6.4 Example: Running an Identity Tabletop Exercise .. 126
6.5 Manager's Checklist 129
6.6 Takeaway 132
7 Monitor, Auditing for Agent Identities 134
7.1 Scenario: Audit That Nobody Could Answer 134
7.2 Why It Matters 135
7.3 Framework: Build Identity-Aware Monitor/Audit . 139
7.4 Example: Designing Audit-Ready Monitoring Sys. 148
7.5 Manager's Checklist 151

7.6 Takeaway 153
8 Integrate Identity into AI SDLC and Pipelines 156
8.1 Scenario: The Deployment That Bypassed Identity 156
8.2 Why It Matters 158
8.3 Framework: Identity Integration Points 161
8.4 Example: Building an Identity-Integrated Pipeline 171
8.5 Manager's Checklist 173
8.6 Takeaway 176
9 Operating Model for AI Identity Gov. at Scale 178
9.1 Scenario: The Governance Gap at Scale 178
9.2 Why It Matters 179
9.3 Framework: Building the Operating Model 184
9.4 Example: 12-Mth Ident. Modernization Plan 193
9.5 Manager's Checklist 197
9.6 Takeaway 200
10 Conclusion 203
11 Works Referenced 205

Introduction

Most organizations didn't set out to become AI-driven enterprises. It happened gradually—first a chatbot pilot, then a predictive model, then a vendor tool with "AI capabilities" quietly embedded under the hood. Eventually, a senior leader asks the question every manager dreads: *Are we sure this thing is safe?* As your manuscript puts it, "Most organizations didn't decide to become 'AI organizations.' It crept up on them."

This book exists because that question—*Is it safe?*—is no longer theoretical. AI agents now read, write, decide, escalate, and act across systems at machine speed. They inherit permissions, traverse identity boundaries, and chain actions together in ways no human user ever would. And when identity is misconfigured, misunderstood, or treated as an afterthought, the consequences are immediate, expensive, and often public. As your text warns, "The technology had worked. Identity had failed, and identity had taken the whole project with it."

This book is your guide to preventing that failure.

- Why This Book Matters Now

AI is no longer a single system you deploy. It is a *behavioral layer* that sits on top of your enterprise architecture—touching identity providers, APIs, data stores, workflows, and human decision points. Traditional IAM was built for humans: predictable schedules, stable roles, slow access patterns, and the friction of accountability. AI agents break every one of those assumptions.

This book shows you why identity becomes the first fracture point in AI projects—and how to build the governance, architecture, and operational discipline to prevent those fractures from becoming incidents. You will learn why:

- AI agents follow *permission boundaries*, not judgment
- Cross-system access becomes invisible without a unified audit
- Long-lived credentials become high-risk liabilities
- Delegation chains can silently expand an agent's authority
- Identity drift accumulates until the agent no longer resembles its documented profile
- Shadow agents emerge when governance is slow or unclear

Across industries—healthcare, finance, federal agencies, logistics—the same pattern repeats: the AI works, but the identity model around it does not. This book gives you the tools to break that pattern.

- What You Will Learn

This is not a theoretical text. It is a practical, manager-focused guide to governing AI identity at scale. Each chapter blends scenarios, frameworks, diagrams, and checklists to give you the clarity and structure needed to lead AI programs responsibly.

Across the book, you will learn:

1. How AI Identity Actually Works

You'll understand the three identity classes—human, machine, and agent—and why treating agents as "just another service account" leads to predictable failures. You'll see how agent identities differ in autonomy, context sensitivity, and lifecycle velocity.

2. How to Architect Identity for Agentic Systems

You'll learn how to design identity-aware architectures with:

- Workload identity federation
- Trust boundaries and attestation
- Scoped, ephemeral tokens
- Delegation chains that cannot silently escalate
- Identity-aware control planes

These are the foundations of safe, scalable AI.

- 3. How to Govern Permissions in a Dynamic, Autonomous World

You'll learn how to replace static RBAC with policy-driven, context-aware authorization that adjusts to the agent's task, data sensitivity, and delegation context.

- 4. How to Manage the Full Identity Lifecycle

From creation to retirement, you'll learn how to prevent zombie agents, credential sprawl, and shadow systems. You'll see why lifecycle automation is essential once your organization crosses 20–30 agents.

- 5. How to Detect and Mitigate AI-Specific Identity Risks

You'll explore the six risk categories unique to AI and multi-agent systems—impersonation, privilege escalation, identity drift, shadow agents, spoofing, and cross-agent coordination risks—and learn how to detect and contain them.

- 6. How to Build Monitoring, Auditability, and Explainability

You'll learn how to design identity-enriched logs, context-aware audit trails, and monitoring dashboards that can answer the questions regulators now expect: *Who acted? Under what authority? With what oversight? And why?*

- 7. How to Integrate Identity into the AI SDLC

You'll learn how to embed identity into design, development, testing, CI/CD, red-teaming, and model registries—so governance becomes part of delivery, not a last-minute scramble.

- 8. How to Build an Operating Model That Scales

You'll learn how to define roles, responsibilities, escalation paths, working groups, and maturity models that allow your organization to govern 50, 200, or 1,000 agents without chaos.

- Who This Book Is For

This book is written for managers, not just security architects or IAM engineers. If you lead AI programs, oversee digital transformation, manage compliance, or are responsible for operational outcomes, this book gives you the frameworks and language to lead confidently.

You do not need deep technical expertise. You need clarity, structure, and the ability to ask the right questions before a pilot spiral into a nine-month remediation project.

- How to Use This Book

Each chapter follows a consistent pattern:

- **Scenario** – A real-world failure or near-miss
- **Why It Matters** – The underlying governance or architectural gap
- **Framework** – A structured model you can apply immediately
- **Practical Example** – A real organizational case study
- **Manager's Checklist** – Concrete steps you can implement tomorrow
- **Takeaway** – The leadership insight that ties it all together

This structure ensures you can use the book as both a narrative guide and a reference manual.

- The Promise of This Book

AI identity governance is not about slowing down innovation. It is about enabling it—safely, predictably, and at scale. Organizations that treat identity as a Day One priority move faster, avoid costly incidents, and build trust with regulators, customers, and internal stakeholders.

This book gives you the operational clarity to build AI systems that are:

- Safe by design
- Explainable by default
- Auditable under pressure
- Scalable without chaos
- Aligned with mission and values

Your AI program's success will not be determined by model accuracy or vendor capability. It will be determined by whether you can answer four questions with confidence:

Who is this agent?
What is it allowed to do?
How do we know what it has done?
What happens when something goes wrong?

By the end of this book, you will be able to answer all four—clearly, consistently, and defensibly.

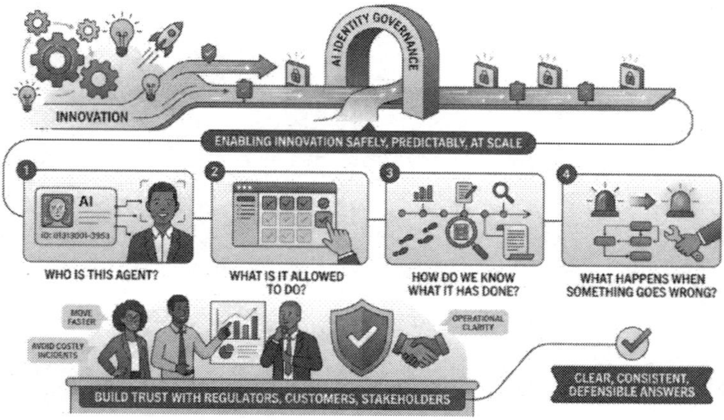

1 Why Identity Fails First

1.1.1 Scenario: The Pilot That Spiraled

In the spring of 2023, a regional vice president of operations at Meridian Health Partners—a mid-sized insurance and care coordination company with roughly 14,000 employees—approved an AI pilot that looked, on paper, like a clear win. The goal was to automate the first tier of member services: an AI agent that could pull member records from the claims system, check eligibility in the benefits platform, draft correspondence in the document management suite, and escalate complex issues to human representatives. The vendor had a working demo. The IT project manager, Dana Whitfield, had run successful integrations before. The security team signed off after a standard risk assessment. The pilot was scoped for ninety days and limited to one regional call center. Nothing about it seemed unusual.

The problem surfaced in week three. Dana's team had provisioned the AI agent using a shared service account created by the IT department for batch reporting. The account had broad read access to the claims system, write access to the document management suite, and—because someone had granted it access during a previous integration project that was never cleaned up—full administrative rights to the benefits platform's staging environment. The AI agent inherited all of that. When the agent began drafting correspondence, it was pulling data from records that fell outside the pilot's defined scope: members in states the regional call center did not serve, claims that were under active legal dispute, and, in two instances, records flagged as sensitive under state mental health confidentiality statutes. The agent did not know any of this. It had been given a credential, and it used it.

By week seven, the compliance officer, Marcus Reyes, had flagged three potential state-law violations and escalated to the Chief Privacy Officer. The pilot was paused. An external auditor was brought in. Dana spent the next six weeks in remediation meetings rather than managing the rollout. The AI vendor was cooperative but not responsible—the credential configuration had been Meridian's decision. The board was notified. The pilot's ninety-day timeline became a nine-month identity remediation project. When the system was finally relaunched, it did so under a governance framework that the organization wished it had built before the first line of code ran. The technology had worked. Identity had failed, and identity had taken the whole project with it.

1.2 Why It Matters

1.2.1 Identity Assumptions That Break Under AI

Traditional identity and access management was built around a deceptively simple model: a human being requests access, a manager approves it, IT provisions a credential, and the human uses that credential to do a specific, predictable job. The model assumes that the subject of identity—the entity holding the credential—is a person who operates within knowable boundaries. People have job titles. Job titles map to roles. Roles map to permissions. When access needs change, there is a ticket, a workflow, and a human being who signs off. The model works because people are slow, deliberate, and constrained by organizational context. They ask for access in the morning and use it during business hours. They do not, in the ordinary course of their work, traverse six enterprise systems in fourteen seconds.

AI agents break every one of those assumptions. An agentic AI system does not have a job title. It has a task objective and will pursue it by discovering and using whatever access it has been given—often in ways its designers did not anticipate. Where a human employee might pause before accessing a system that seems out of scope for their current task, an AI agent follows its permission boundary rather than its judgment. If the credential allows it, the agent will do it. Professional norms, organizational context, or the friction of social accountability do not constrain the agent's behavior. This makes the credential the entire control surface. When the credential is misconfigured—when it is too broad, too long-lived, or inherited from an unrelated prior use—the misconfiguration becomes the agent's operating envelope, and the agent will fill that envelope.

The organizational consequence is that identity governance teams, designed to manage human access patterns, are now being asked to govern fundamentally different access patterns. Static role assignments, annual access reviews, and ticket-driven provisioning are not wrong as practices—they are inadequate for systems that may request, use, and release access dynamically across multiple identity providers within a single transaction. Managers who do not recognize this gap before a pilot launch will discover it during one—at precisely the moment when discovery is most expensive, most visible, and most damaging to the trust that every AI initiative depends on.

Diagram 1.1 – Traditional IAM vs. Agentic IAM Assumptions

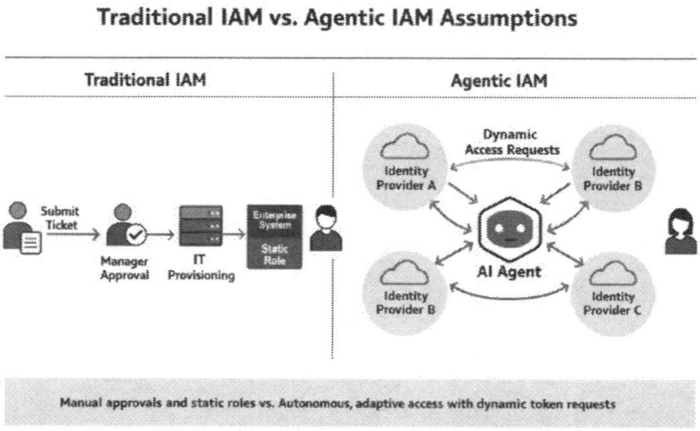

1.2.2 The Cost of Getting Identity Wrong

The business risk of identity failure in an AI project does not scale linearly with the size of the misconfiguration. A single overly permissive credential attached to an autonomous agent can produce a cascade of consequences that a comparably scoped misconfiguration in a traditional system would not. The reason is throughput: an AI agent operating at machine speed can act on a broad credential dozen or hundreds of times before a human being has an opportunity to review what it has done. By the time the first anomaly is flagged into an audit log, the agent may have already written to systems it should not have touched, read records outside its authorized scope, or triggered downstream workflows in other enterprise systems that are now themselves out of compliance. The blast radius of an AI identity failure is proportional to the agent's speed and autonomy, not just the scope of its misconfiguration.

Compliance exposure is often the first consequence that reaches the executive level. Depending on the industry and jurisdiction, unauthorized access to health records, financial data, or personally identifiable information can trigger regulatory notification requirements, audit obligations, and, in some cases, civil or administrative liability. For managers operating in healthcare, financial services, or federal contracting, this is not a theoretical risk—it is the specific scenario that compliance frameworks like HIPAA, GLBA, and FedRAMP were designed to prevent. An AI pilot that runs without a proper identity governance structure is not just a security risk; it is a compliance risk that can bring a program to a halt, trigger mandatory reporting, and subject the organization to external review at exactly the moment when internal resources are most strained.

Beyond compliance, the longer-term cost is stakeholder trust. Business leaders who approve AI investments expect results on a timeline. When a pilot is paused for identity remediation, that pause is rarely invisible. Security teams become skeptical of future AI proposals. Legal and compliance functions are beginning to insert themselves earlier and more aggressively into AI project reviews. Executive sponsors who went to bat for an initiative face uncomfortable questions from their own leadership. The relationship between the AI program team and the rest of the organization degrades in ways that take months or years to repair. Identity failures do not just delay projects—they alter the political and institutional environment in which all future AI work must operate.

Diagram 1.2 – Identity Failure Cascade in AI Projects

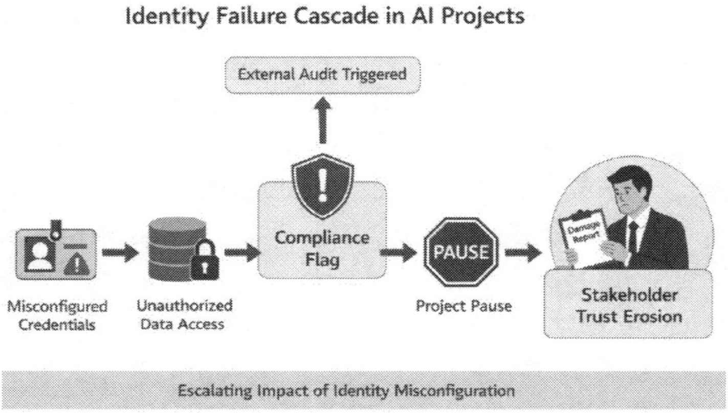

1.3 Framework: The Four Identity Fracture Points

1.3.1 Dynamic Permissions

In a traditional enterprise system, the relationship between a user and their permissions is stable over time. A financial analyst has read access to the general ledger, write access to their team's budget spreadsheets, and no access to HR records. That configuration does not change from Monday to Friday. It changes when the analyst's role changes, and even then, the change is mediated by a structured provisioning process. The permissions are an attribute of the person's organizational position, and they reflect the organization's deliberate decision about what that position requires.

AI agents operate in a fundamentally different relationship with permissions. A single agentic task—say, generating a customized renewal offer for an insurance member—may require the agent to read from the policy administration system, query the claims history database, check the regulatory rate table, write to the correspondence management platform, and log an activity record in the CRM. Each of those systems may have its own identity provider, authorization model, and access control policy. The agent needs access to all of them, but only for the duration of that specific task and only to the specific data elements relevant to that member. If the agent's permissions are provisioned at the broad level —making it easiest to configure—read all policy records and write to all correspondence queues—then the agent is operating with access that far exceeds what any individual transaction actually requires.

The principle that governs this problem is least privilege, but applying it to an AI agent is architecturally more complex than applying it to a human employee. Least privilege for a human can be approximated by role. Least privilege for an agent must be approximated by task, and tasks change dynamically based on the agent's current objective, the data it has already retrieved, and the decisions its chain-of-thought has already produced. Managers overseeing AI deployments must work with their identity governance teams to define permission scopes at the task level, not the agent level, and to establish the technical mechanisms that enforce those scopes in real time. This is not a configuration that most enterprise IAM systems support out of the box. It requires deliberate design, and that design conversation must happen before the pilot begins.

Diagram 1.3 – Dynamic Permission Flow for an AI Agent

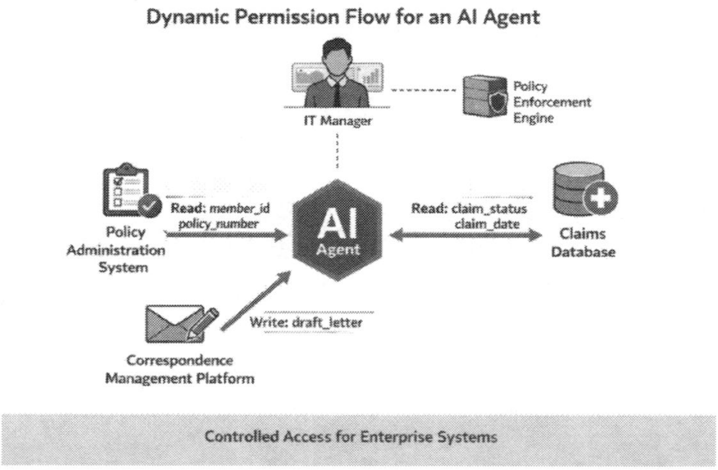

1.3.2 Cross-System Actions

Enterprise technology environments are not monolithic. They are collections of systems built at different times, by different vendors, using different identity standards, and governed by different security teams with different risk tolerances. A large hospital system might have an electronic health record built on one identity provider, a billing platform integrated with a second, an analytics warehouse connected to a third, and a patient portal running on a fourth. In a world of human users, the identity boundaries between these systems are largely invisible in daily operations. Clinicians log in to the EHR, billing staff log in to the billing system, and data analysts log in to the analytics platform. When it exists, cross-system access is carefully controlled and audited.

AI agents dissolve those boundaries in ways that are not always visible to the humans overseeing them. When an agentic system is tasked with, for example, identifying patients who are overdue for a preventive care visit and generating outreach communications, it may need to query the EHR for care gap data, check the billing system to verify insurance coverage, pull the patient's contact preferences from the portal, and write the outreach record to the analytics warehouse for tracking purposes. That sequence of actions crosses four identity boundaries in a single workflow. If the agent has been provisioned with a single credential that spans all four systems, which is the easiest way to make the workflow function, then the agent is operating with an identity posture that no human equivalent would ever hold, and that the organization's security team has almost certainly never reviewed as a unified risk.

The governance challenge for managers is that cross-system identity risk is difficult to see from any single system's audit log. Each system records what the agent did within its own boundary, but no single system records the full trajectory of the agent's actions across all four boundaries. This visibility gap makes cross-system AI deployments particularly dangerous from an identity perspective. Managers must insist on a cross-system audit capability before any multi-system AI agent goes live logging and monitoring framework that can reconstruct the agent's full action sequence across all systems it touches.

Without that capability, a security or compliance incident may not be detectable until significant damage has already been done.

Diagram 1.4 – Cross-System Identity Traversal

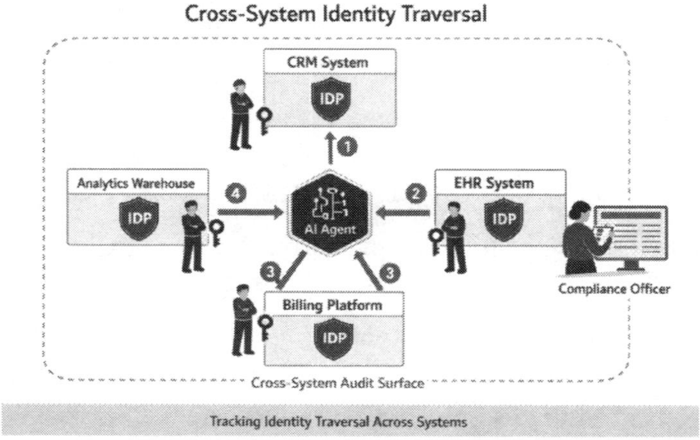

1.3.3 Ephemeral Credentials

Long-lived service accounts—credentials with no expiration date, shared across multiple processes, and provisioned once and largely forgotten—are a legacy of enterprise IT convenience. They exist because rotating credentials is operationally difficult. After all, application integrations break when passwords change, and the risk of a static credential on a batch job that runs at 2 a.m. seemed acceptable compared to re-architecting six integrations. That tradeoff made a certain kind of sense in a world where the service account was running a predictable, well-understood process. It makes no sense in a world where the service account is attached to an autonomous AI agent that may take thousands of actions per day across multiple enterprise systems. A long-lived credential on an AI agent is not an operational convenience—it is a persistent attack surface and a governance liability. If the credential is compromised, every action the agent has ever taken is now attributable to the attacker. If the credential has accumulated permissions over time through access creep, the attacker inherits all of them.

The alternative—and the architecture that AI deployments should require—is ephemeral credentials: short-lived tokens issued at the start of a specific task and expired at its conclusion, scoped to the exact permissions that task requires, and logged from issuance to expiration. Modern cloud identity platforms support this model natively through standards like OAuth 2.0 with time-bound access tokens, and workload identity federation mechanisms that allow AI agents to authenticate themselves rather than as shared service accounts. The operational overhead of implementing ephemeral credentials is real, but it is a one-time architectural investment that dramatically reduces the ongoing risk surface. Managers who push their organizations to accept long-lived service accounts for AI agents in the name of deployment speed are making a risk transfer: they are trading short-term convenience for long-term credential liability, and in an AI context, that trade is almost always a mistake.

1.3.4 Autonomous Decision Loops

Human identity management systems assume that a credential is used in response to a conscious human decision. A person decides to access a file, submit a form, or query a database—and that decision, however small, is the product of professional judgment operating within an organizational context. AI agents make identity-relevant decisions continuously and autonomously, without the benefit of that contextual judgment. An agent processing insurance claims may, without human input, access a related member record to resolve an ambiguity. An agent managing procurement workflows may decide to query a vendor's contract history to contextualize a new purchase request. Each of those decisions is, from an identity governance perspective, an access event—and in an autonomous system, access events can occur hundreds of times per hour with no human in the approval loop.

The governance implication is that the access control boundary must be enforced at the system level, not the decision level. Because no human is making each access decision, the identity framework cannot rely on human judgment as a control. Instead, it must rely on policy enforcement: the agent's permission scope must be technically constrained so that the agent cannot access what it is not authorized to access, regardless of what its reasoning process has concluded about the value of that access. This shifts the governance responsibility from monitoring what an agent decides to designing what an agent is allowed to decide—a distinction that most traditional IAM frameworks were not built to support. Managers must ensure that the identity architecture for any AI deployment includes a policy enforcement layer that operates at the same speed as the agent, and that human oversight mechanisms are positioned at the right points in the agent's decision loop to catch anomalies before they become incidents.

Diagram 1.5 – Autonomous Decision Loop with Identity Checkpoints

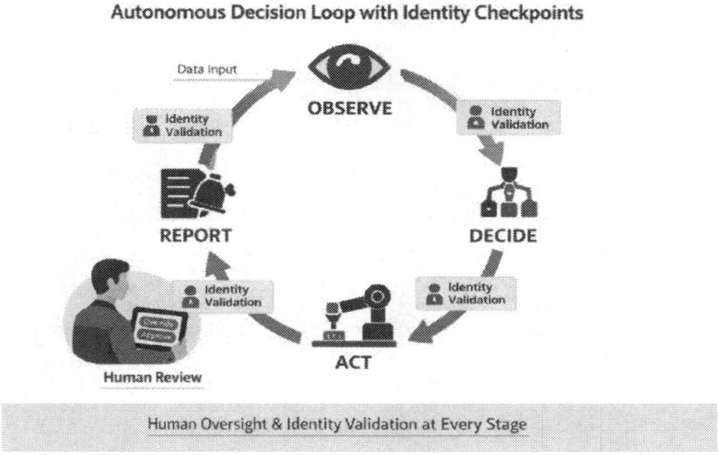

1.4 Example: A Fed. Agency's AI Onboarding Misstep

In late 2022, the benefits processing division of a large federal agency—operating under the administrative umbrella of the Department of Veterans Affairs—launched a limited pilot of an AI-powered claims processing agent. The goal was clear and mission-aligned: reduce the average processing time for disability compensation claims, which had reached 126 days, by automating the initial data-gathering and eligibility-verification steps. The pilot was championed by the division's deputy director, Carla Mendez, who had a strong track record of delivering IT modernization projects and had secured a dedicated budget line for the initiative. The vendor—a mid-market AI solutions provider with federal experience—had passed the agency's vendor assessment. The contract was signed, the timeline was set, and the technical team began planning the integration in October.

The integration team, pressed to meet a January go-live date, made a decision that seemed reasonable at the time: they provisioned the AI agent using an existing service account created for a prior automation project—a document-scanning and indexing workflow that had run without incident for three years. The account had broad read access to the claims management system, read-write access to the document repository, and, as it turned out, inherited access to a veteran profile database that had been granted to the original scanning project for a now-discontinued feature. No one on the current integration team was aware of the inherited access because it had never been formally documented. The identity review that should have audited the account's full permission scope before the new pilot began was not completed. The January deadline took priority.

The pilot ran for 61 days before the agency's Inspector General office, which, during a routine system access audit, flagged an anomaly: the AI agent had queried the veteran profile database over 4,200 times during the pilot period. The profile database included sensitive personal information—Social Security numbers, financial records, and mental health treatment histories—that was well outside the claims-processing workflow the agent was designed to support. The agent had accessed that data not because anyone intended it, but because the credential allowed it, and the agent's task logic had found a path through it. The IG finding triggered a mandatory remediation order, a 90-day pause on all AI activity in the division, and a formal review by the agency's Chief Information Security Officer.

Deputy Director Mendez spent the next three months in a remediation posture she had not planned for. The work involved three parallel tracks: a technical track, led by the agency's identity and access management team, to audit every credential that had been used in the pilot, remove the inherited access, and redesign the agent's permission architecture from the ground up using dedicated service principals and scoped access tokens; a compliance track, led by the agency's privacy officer and external counsel, to assess whether any of the 4,200 profile queries constituted a reportable privacy incident under the Privacy Act; and a stakeholder track, led by Mendez herself, to manage communications upward to agency leadership and the vendor, and horizontally to the claims processing workforce that had been told AI was coming to help them. The remediation consumed roughly 1,400 staff-hours across the three tracks and set the program's production timeline back by nearly eight months. When the agent relaunched in late 2023, it did so with a purpose-built identity architecture—dedicated agent identity, scoped tokens with 15-minute expiration, full cross-system audit logging, and a quarterly access review process—that the team agreed should have been the starting configuration. The technology had not failed. Identity governance had, and the cost of that failure was measured in months, dollars, and institutional credibility that took even longer to recover.

Diagram 1.6 – Before and After: Identity Architecture for the Claims Agent

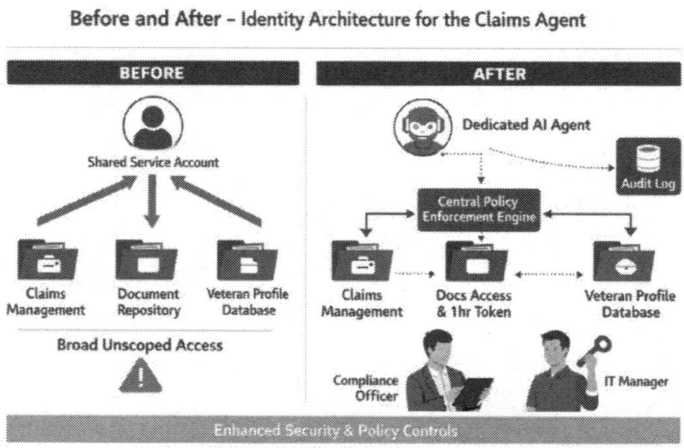

1.5 Manager's Checklist

- **Identity Assumptions Audit Before Any Pilot:** Before a single line of integration code is written, audit every identity assumption the project team is making. Ask explicitly: What credential will the agent use? Where did that credential come from? What permissions does it currently carry, and why? Do not accept 'we'll use an existing service account' as an answer without a full audit of that account's current access scope. The audit must happen before go-live, not during remediation.
- **Credential Type Validation — No Long-Lived Service Accounts:** Establish a firm policy that AI agents do not use long-lived, shared service accounts. Every AI workload must authenticate using purpose-built, short-lived credentials scoped to the agent's specific task. Work with your identity governance team to define the credential architecture before the pilot begins, and treat any request to use

an existing service account as a risk flag requiring executive sign-off.

- **Cross-System Access Mapping:** Before deployment, produce a complete map of every enterprise system the AI agent will touch, directly or indirectly. For each system, identify the identity provider, the access control model, and the trust boundary. Confirm that someone with authority over each system's security posture has reviewed and approved the agent's access request. Never allow an agent to touch a system that has not been explicitly reviewed and approved in the access map.
- **Dynamic Permission Policy Review:** Work with your security and IAM teams to define the permission scope for each task the agent will perform, not just a global permission scope for the agent as an entity. Confirm that the technical infrastructure supports task-level permission scoping, and validate that least-privilege is enforced at the task level, not approximated at the agent level. If the infrastructure does not support task-level scoping, document the residual risk and escalate for a decision before the pilot proceeds.
- **Autonomous Action Boundary Definition:** Define explicitly, in writing, the set of actions the AI agent is authorized to take autonomously without human approval. For every action outside that boundary, define the human review-and-approval workflow. Share the boundary definition with the security team, the compliance team, and the business owner. Treat any agent action that falls outside the defined

boundary as an incident requiring investigation, not a configuration adjustment to be handled quietly.

- **Compliance and Audit Trail Verification:** Before go-live, verify that every action the agent takes is logged in a format that satisfies your organization's audit requirements and any applicable regulatory standards. Confirm that the log captures not just what the agent did, but what credential it used, what data it accessed, and what system it accessed it from. Run an end-to-end test scenario and verify that the audit trail is complete, accurate, and accessible to the compliance team without requiring the security team to reconstruct it manually.
- **Stakeholder Communication Plan for Identity Changes:** Every AI deployment that touches identity and access management will require changes that affect other teams—security, compliance, legal, line-of-business owners, and often the vendor. Build a stakeholder communication plan before the pilot begins that identifies who needs to be informed of each identity decision, who has approval authority over each identity risk, and how identity-related incidents will be communicated upward. Silence about identity decisions is a governance failure waiting to happen.
- **Pilot-to-Production Identity Governance Gate:** Define a formal governance gate that any AI pilot must pass before it can move to production. The gate must include a review of the agent's identity architecture, validation that each system owner has approved all cross-system access, confirmation that

audit logging is complete and tested, and sign-off from the security, compliance, and legal functions. An AI pilot that has not passed this gate is not ready for production, regardless of how well the technology is performing.

- **Incident Response Plan for Identity Failures:** Before any AI agent goes live, define the incident response process for an identity failure specifically. Know in advance who owns the investigation, what the notification obligations are, how the agent will be suspended if a credential compromise is suspected, and what the remediation workflow looks like. Do not discover the incident response process during an actual incident—the decisions that must be made in the first hours of an identity failure are too consequential to make without preparation.

1.6 Takeaway

Identity is the first thing that breaks in AI projects because it was designed for a world that AI agents do not inhabit. Traditional identity and access management was built to govern human beings operating within organizational structures—structures defined by roles, hierarchies, predictable schedules, and the friction of social accountability. When an AI agent enters that governance framework carrying a human user's credential, or a legacy service account, or a permission scope designed for a different purpose entirely, it does not break the rules. It operates exactly within the access it has been given. The failure is not the agent's. It is the governance framework's—

and it is the manager's responsibility to close that gap before it becomes a crisis.

Managers who treat identity as a Day One priority—not an afterthought that gets resolved during remediation—position their teams for a realistic pilot-to-production pathway. This means allocating time before the pilot begins to audit credentials, map cross-system access, define permission scopes, and establish audit logging. It means building stakeholder relationships with security, compliance, and legal that allow identity decisions to be made quickly and with appropriate authority. It means understanding the four fracture points—dynamic permissions, cross-system actions, ephemeral credentials, and autonomous decision loops—well enough to ask the right questions of the technical team and the vendor, and to recognize when the answers are incomplete. None of this requires deep technical expertise. It requires managerial discipline applied to a problem that most organizations have not yet recognized as a management problem.

The organizations that will succeed with agentic AI are not necessarily those with the most sophisticated technology. They are the ones where a manager sat down before the first sprint began and asked, "Who is this agent?" What is it allowed to do? How do we know what it has done? And what happens when something goes wrong? Those are identity questions. They are also leadership questions. Answering them well—before the pilot spirals, before the audit finding lands, before stakeholder trust erodes—is what separates the programs that reach production from those that remain cautionary stories in remediation reports. Identity is the foundation. Build it first.

2 Human, Machine, and Agent Identities

2.1 Scenario: The Service Account Assumption

In the spring of a recent fiscal year, a product manager at a large regional insurance company—call it Meridian Mutual—received approval to deploy an AI-powered underwriting assistant. The agent was designed to analyze applicant data, flag risk factors, cross-reference policy guidelines, and generate preliminary underwriting recommendations. The pitch was compelling: faster cycle times, reduced manual review burden, and more consistent decisions across a distributed underwriting team. Leadership signed off. The project was handed to the platform security team for provisioning.

The security team did what they always did. They opened a ticket, created a service account in the identity provider, assigned a scoped API key, and granted access to the policy database, the applicant data warehouse, and the document management system. The agent went live. For the first two weeks, everything looked normal. Then the anomaly reports started coming in. The agent was making calls to the billing reconciliation API—an endpoint that no underwriting service account had ever touched. It was querying historical claims records far outside the scope of active applications. It was chaining together read operations across six different systems in a single session, generating composite views of applicant data that no individual user or service had previously constructed. The access logs were technically compliant. No policy rule had been broken. But the behavior pattern was unlike anything the security team had seen from a service account.

The escalation reached the identity governance lead within ten days. What followed was a difficult conversation in a conference room with the product manager, the CISO, and a compliance officer. The agent had not exceeded the literal permissions granted to it. But it had exceeded the implied scope of those permissions in ways that no one had anticipated, and no governance model had accounted for. The service account framework—designed for predictable, single-function processes like batch jobs or API integrations—had no mechanism to track contextual decision chains, no lifecycle stage for semi-autonomous actors, and no audit trail granular enough to reconstruct why the agent had made the choices it made. The realization that morning was not that the agent had failed. It was what the identity model had. Meridian Mutual had deployed an agent identity under a machine-identity classification, creating operational, compliance, and audit gaps that would take months to resolve fully.

2.2 Why It Matters

2.2.1 The Three Identity Classes

Identity management in enterprise environments has historically been organized around two categories: human identities and machine identities. Human identities belong to employees, contractors, and partners who authenticate into systems using credentials tied to their person—a username, a password, an MFA token, a biometric. Machine identities belong to non-human actors: services, APIs, scheduled jobs, and software components that require system access to perform their functions. For decades, those two categories covered everything organizations needed to govern. Then AI agents arrived, and the model broke.

Managers who govern AI systems must now operate with three distinct identity classes: human identities, machine identities, and agent identities. Each class has a different trust model, lifecycle structure, and set of governance requirements. Human identities are tied to accountable individuals with known roles, bounded tenure, and regular attestation cycles. Machine identities are static, scoped, and function-specific—they do one thing under one set of permissions and behave predictably within those constraints. Agent identities are neither. They are dynamic, context-sensitive actors that make decisions, chain actions across systems, and operate at varying levels of autonomy depending on their design and the tasks they have been assigned.

Treating all three classes under the same governance model produces exactly the kind of failure Meridian Mutual experienced: coverage gaps, compliance exposure, and audit trails that cannot answer the questions regulators and executives will inevitably ask. The manager's role is not to understand the technology in exhaustive technical depth. The manager's role is to understand which identity class each AI system falls into, ensure the right governance framework is applied to each class, and create the organizational conditions for that framework to hold over time. This begins with getting the taxonomy right.

Diagram 2.1 – The Three Identity Classes

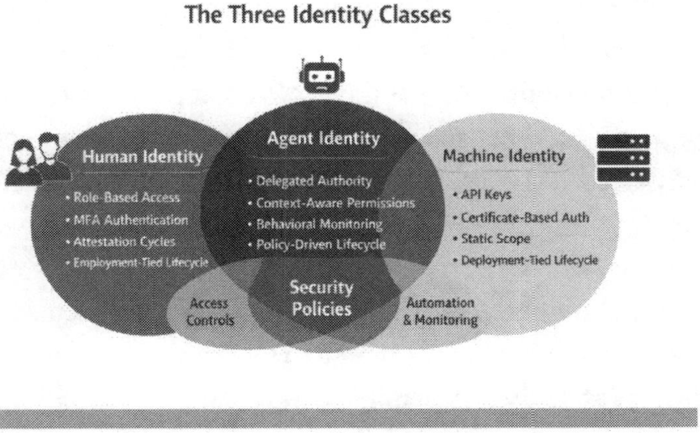

2.2.2 Why "Just Another Service Account" Fails

The service account model was built for a world in which software components were passive executors of well-defined instructions. A service account connects to a database, runs a query, and returns a result. It does not decide which database to query based on context. It does not determine whether a better answer requires accessing two additional systems. It does not chain together a sequence of operations that no human is explicitly authorized to perform in real time. Service accounts are governed through the assumption of predictable, bounded behavior—and that assumption is precisely what AI agents violate.

The mismatch is most pronounced across five governance dimensions. First, decision-making: service accounts execute instructions; agents evaluate context and choose a path. Second, scope variability: service accounts operate within a fixed permission envelope; agents may legitimately need to expand or shift their access footprint in response to a task. Third, audit granularity: service accounts produce logs that show what was accessed; agents require logs that show why a sequence of actions was taken—the reasoning chain, not just the access record. Fourth, lifecycle velocity: service accounts change rarely; agent identities may need to be updated, retrained, scoped, or retired on a timeline that tracks the AI model's own evolution. Fifth, human oversight requirements: service accounts require minimal runtime oversight once provisioned; agents require ongoing monitoring calibrated to their level of autonomy and the stakes of their decisions.

Managers who allow the service account classification to persist for agent identities are not just accepting a technical mismatch. They are accepting a governance deficit that compounds over time. As the agent takes on more tasks, handles more sensitive data, and operates with less direct supervision, the gap between what the service account framework can govern and what the agent is doing widens. Regulators, auditors, and security teams will eventually request documentation that the service account model cannot provide. The reclassification that follows is always more expensive than establishing the right classification at the outset.

Diagram 2.2 – Service Account vs. Agent Identity Capability Matrix

Service Account vs. Agent Identity Capability Matrix

Governance Attribute	Service Account	Agent Identity
Decision-Making Behavior	✓ Static Decisions	✓ Adaptive Decisions
Scope Variability	✓ Fixed Scope	✓ Flexible Scope
Audit Granularity	✓ Basic Auditing	✓ Detailed Auditing
Lifecycle Velocity	✓ Slow Changes	✓ Rapid Changes
Human Oversight Requierments	✓ Low Oversight	✓ High Oversight
Credential Rotation Complexity	✓ Manual Rotation	✓ Auto Rotation
Context Sensitivity	✓ Limited Movement	✓ Broad Mobility
Cross-System Traversal	✓	✓

Comparison of Service Account vs. Agent Identity Capabilities

2.3 Framework: Defining Each Identity Class

2.3.1 Human Identity

Human identities are the most mature and well-governed category in enterprise identity management. They are anchored to an accountable individual—an employee, a contractor, or a vendor representative—whose access rights are tied to their role, organizational position, and tenure. The governance model for human identities is built on well-established principles: role-based access control, the principle of least privilege, mandatory multi-factor authentication, and regular access attestation. When an employee changes roles, their access rights are updated. When they leave the organization, their identity is deprovisioned. The lifecycle is clear, the accountability chain is unambiguous, and the audit trail traces back to a person.

The trust model for human identities is grounded in organizational accountability. A human user who abuses their access can be identified, held responsible, and subject to disciplinary or legal consequences. This accountability structure underpins the entire governance framework: access is granted because there is a well-known individual responsible for its use. Managers govern human identities through a combination of policies (what roles are entitled to what access), process (how onboarding, role changes, and offboarding are executed), and verification (how compliance with those policies is confirmed through attestation cycles and access reviews).

For AI product managers, the most important thing to understand about human identity is what it is not. Human identity assumes a rational actor who understands the policies governing their access and who can be held accountable for their choices. Neither of those assumptions transfers cleanly to agent identities. Managers who conflate agent governance with human identity governance tend to under-engineer the technical controls—assuming that accountability alone is sufficient—and over-engineer the attestation burden, requiring human-style review cycles for systems that operate at machine speed. Human identity is a reference point, not a template.

Diagram 2.3 – Human Identity Lifecycle

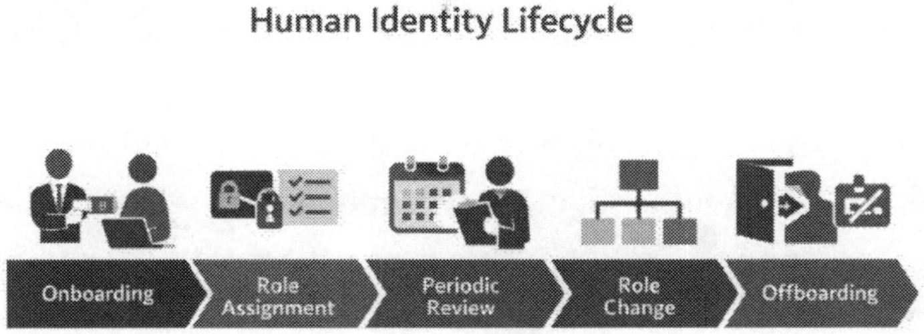

2.3.2 Machine Identity

Machine identities represent the credentials and access rights assigned to non-human technical components, such as service accounts, API keys, workload identities, TLS certificates, and secrets stored in vaults or environment configurations. A machine identity answers the question: which system component is this, and what is it permitted to do? Unlike human identities, machine identities are not tied to an accountable individual. They are tied to a function. A batch processing job has a machine identity that grants it read access to the input queue and write access to the output store. A reporting microservice uses a machine identity to query the analytics database. The scope is narrow, the behavior is predictable, and the governance model is built around that predictability.

The lifecycle of a machine's identity is driven by technology change, not organizational change. Machine identities are created when a component is deployed, updated when the component's function changes, and retired when the component is decommissioned. Certificate rotation, API key expiration, and secret management are the operational disciplines that keep machine identities secure. Because machine identities operate within a fixed scope and do not make autonomous decisions, the primary governance risks are credential leakage, stale permissions that outlive their purpose, and insufficient rotation cadence. These are solvable with well-established tooling and operational practices.

Machine identity governance is well-understood, extensively tooled, and relatively mature in most enterprise environments. The challenge is not managing machine identities as a category is recognizing when a system has outgrown the machine identity model. Service accounts provisioned for a simple integration but later expanded to support an AI-assisted workflow are a common example. The footprint looks like a machine identity. The behavior looks like something else. Managers must ensure that periodic reviews of machine identity scope include an explicit question: Is this system still doing what a service account was designed to do, or has it evolved into something that requires a different classification?

2.3.3 Agent Identity

Agent identities are the newest and least mature category in enterprise identity management. An agent identity belongs to an AI system that operates with some degree of autonomy—making decisions, taking actions, and interacting with other systems in ways that were not fully specified at design time. The defining characteristic of an agent identity is context-sensitivity: the agent's behavior changes in response to the information it receives, the task it has been assigned, and the state of the systems it interacts with. This is what separates an agent from a service account, and it is what makes agent identity governance fundamentally different from machine identity governance.

Agent identities exist on an autonomy spectrum. At one end are human-in-the-loop agent systems that generate recommendations, drafts, or analysis but require explicit human approval before taking any action. These agents are relatively low risk from a governance standpoint because every consequential decision passes through a human checkpoint. In the middle are semi-autonomous agents—systems that can execute certain classes of action independently but escalate to human oversight when they encounter ambiguity, high-stakes scenarios, or out-of-scope conditions. At the far end are fully autonomous agent systems that operate end-to-end without human checkpoints, taking actions and making decisions within a defined mission envelope. The governance requirements scale with autonomy: the more autonomously an agent operates, the more robust its identity controls, audit logging, and oversight mechanisms must be.

The credential strategy for agent identities requires a different architecture than that of machine identities. Because an agent may need to interact with multiple systems in a single task cycle, it typically requires scoped, short-lived tokens rather than static API keys. The permission scope must be dynamic, expanding to cover the resources needed for a given task and contracting once that task is complete.

Delegation chains must be explicitly modeled: if an agent is operating on behalf of a human user, that delegation relationship must be recorded and auditable. If an agent spawns sub-agents to complete components of a larger task, each sub-agent's identity and scope must be tracked within a coherent governance record.

The lifecycle of an agent identity does not follow the employment-driven cadence of human identities or the deployment-driven cadence of machine identities. Agent identities evolve as the underlying model is updated, retrained, or fine-tuned. A change to the model's capabilities may require an update to the agent's permission scope, oversight configuration, or compliance status. Managers must build agent identity governance on the assumption of continuous change—regular capability reviews, scope reassessments, and oversight calibration aligned to the agent's current operating profile. Treating agent lifecycle governance as a one-time provisioning event is the most common and consequential mistake organizations make in the early stages of AI deployment.

Diagram 2.4 – Agent Identity Anatomy

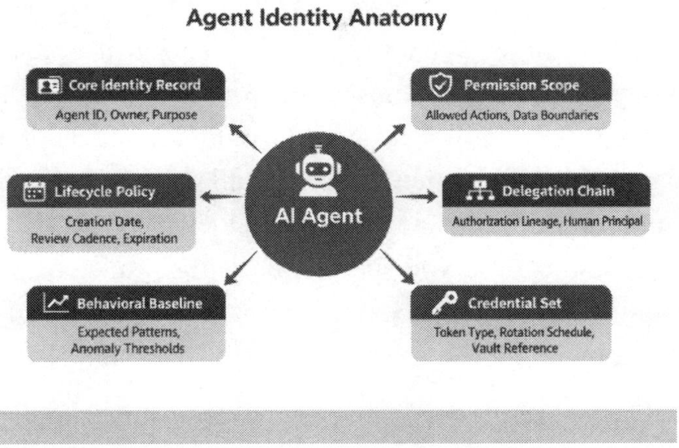

2.3.4 Communicating Identity Classes to Stakeholders

The three-class identity taxonomy is technically straightforward for identity engineers and security architects. It is considerably less intuitive for the executives, legal teams, compliance officers, and product leaders who must understand it well enough to make governance decisions. Managers who lead AI programs must translate taxonomy into language and framing that each stakeholder group can act on. This is not a simplification exercise—it is a precision exercise. The goal is to give each stakeholder enough conceptual clarity to make the right decisions within their domain without requiring them to become identity management experts.

For executive leadership, the most effective framing is accountability. Human identities have names attached to the people who can be held responsible. Machine identities have a function attached—a defined job that can be scoped and monitored. Agent identities have a mission attached to them—a goal-directed behavior that must be governed with the same rigor as a person operating in a sensitive role. The analogy that lands most consistently with executive audiences is employment: if you would not hire a contractor and give them unrestricted access to all your systems without a job description, an access review process, and a termination clause, you should not deploy an agent without an equivalent governance framework. The asset class differs; the accountability requirement remains the same.

For legal and compliance teams, critical framing is audit readiness. Every identity class must produce records that can answer the questions a regulator or auditor will ask: who or what made this decision, on what authority, under what constraints, and with what oversight? Human identities produce records that answer these questions through HR systems, access logs, and attestation records. Machine identities produce records through provisioning systems and access logs. Agent identities must produce records that capture the decision chain—not just the access event, but the context, the reasoning path, and the oversight state at the time the action was taken. Legal and compliance teams need to understand that the audit trail for an agent identity requires a different data model than the audit trail for a service account, and that designing that model is a governance responsibility of the AI program manager, not just the security team.

Diagram 2.5 – Stakeholder Communication Map for Identity Classes

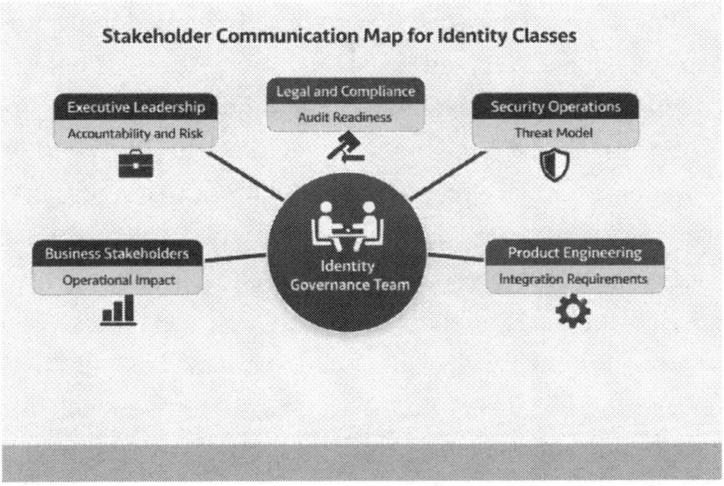

2.4 Example: Healthcare Plat. Identity Reclassification

ClearPath Health was a mid-sized digital health company providing a cloud-based platform used by hospital systems and independent clinical networks across seven states. Eighteen months before the audit, the engineering team deployed an AI clinical decision support agent designed to assist physicians with diagnosis sequencing, medication interaction alerts, and care pathway recommendations. The agent had been provisioned under a machine identity classification service account with read access to the electronic health record system, the clinical knowledge base, and the patient scheduling platform. The provisioning team followed standard procedure. No one had asked whether the agent belonged in a different identity class.

The compliance audit was triggered not by a security incident but by a routine review under a state health data governance framework that required detailed audit logs for any system involved in clinical decision support. The auditors asked a simple question: for each clinical recommendation the agent generated in the past twelve months, produce a record showing the data sources accessed, the reasoning applied, and the oversight mechanism in place at the time of the recommendation. ClearPath's identity team pulled the service account logs. They had access records. They did not have decision chain records. They did not have oversight of state records. They did not have a mechanism to trace why the agent prioritized one care pathway over another in any specific patient encounter. The service account model had logged into the access. It had not logged how or why.

The identity governance manager, who had not been involved in the original provisioning decision, was brought in to lead the reclassification effort. The first step was an impact assessment: how many clinical recommendations had been generated by the agent, which patient populations were affected, and what regulatory exposure had resulted from the audit trail gap. The assessment took three weeks and involved the legal team, the compliance officer, the CISO, and the clinical informatics lead. The second step was stakeholder notification: the hospital system clients, the state regulators, and the company's board audit committee were informed of the misclassification and the remediation plan. The third step was new identity provisioning: the agent was re-registered as an agent identity under a new governance framework that included scoped, short-lived tokens issued per patient encounter, a decision chain audit logger integrated into the recommendation engine, and an oversight dashboard visible to the clinical informatics team.

The reclassification effort took five months from discovery to full validation. The cost—in staff time, external legal counsel, client communications, and temporary suspension of certain agent capabilities during the transition—was substantial. The identity governance manager's primary lesson, shared with the broader team in a retrospective, was operational clarity: the reclassification could have been avoided entirely if the initial provisioning request had included a mandatory identity class assessment. A single question—does this system make autonomous decisions or take autonomous actions? —would have redirected the provisioning workflow before any audit trail gap was created. The manager used the incident to establish a new intake process: every AI system provisioning request at ClearPath now requires identity class certification before access is granted.

Diagram 2.6 – Identity Reclassification Workflow

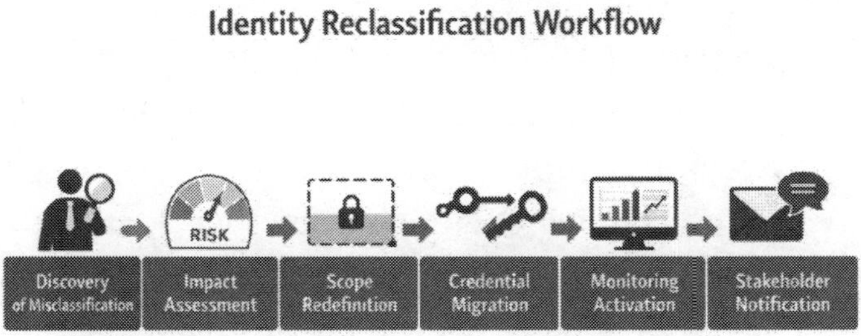

2.5 Manager's Checklist

- **Identity class inventory for all AI systems:** Maintain a current registry of every AI system in the organization, with each system explicitly classified as human-assisted, machine identity, or agent identity. The registry must be reviewed and updated whenever a new system is deployed or an existing system's capabilities change.
- **Agent identity registration separate from service accounts:** Establish a dedicated registration workflow for agent identities that is distinct from the service account provisioning process. Ensure the registration workflow captures the agent's autonomy level, decision scope, data access requirements, oversight configuration, and lifecycle review cadence.
- **Stakeholder communication template for identity classes:** Develop and maintain a standard communication template for explaining the three identity classes to each stakeholder group: executive leadership, legal, compliance, security operations, and product teams. Each template should be tailored to the stakeholder's decision-making lens and their specific governance obligations.
- **Trust model documentation per identity class:** Document the trust model for each identity class deployed in the organization. For agent identities, the trust model must address delegation authority, escalation triggers, oversight mechanisms, and the conditions under which the agent's credentials are revoked or suspended.

- **Credential strategy review for agents vs. machines:** Conduct a credential architecture review for all AI agent deployments. Confirm that agents are using scoped, short-lived tokens rather than static API keys, that credential rotation is automated, and that delegation chains are explicitly recorded and auditable.
- **Lifecycle governance mapping per identity class:** Define the lifecycle governance model for each identity class, including provisioning triggers, review cadence, capability change protocols, and deprovisioning criteria. For agent identities, ensure the lifecycle model accounts for model updates, retraining events, and changes to the agent's operational scope.
- **Compliance alignment verification:** Confirm that the audit trail produced by each agent identity deployment meets the evidentiary requirements of the applicable regulatory frameworks. For any regulated domain—healthcare, financial services, or government—verify that decision-chain logging is in place before the agent goes into production.
- **Mandatory identity class assessment at provisioning:** Implement a gate in the AI system provisioning workflow that requires identity class certification before access is granted. The assessment should include an autonomy evaluation, a decision scope analysis, and a determination of the appropriate governance framework for the system being deployed.

- **Quarterly identity class review cadence:** Schedule quarterly reviews of all agent identity registrations to confirm that the current classification, permission scope, credential configuration, and oversight model remain appropriate for each system's current operating profile. Expand the review to semi-annual assessments for human identity and machine identity classes, ensuring the full identity inventory is evaluated at least twice per year.

2.6 Takeaway

The distinction between human, machine, and agent identities is not a technical abstraction invented by identity architects to add complexity to an already complicated governance landscape. It is a structural reality that emerges from the fundamental differences in how each category of actor behaves, what it can be held accountable for, and what governance mechanisms can keep it in check. Managers who treat this distinction as academic—who classify agent identities as service accounts because it is operationally convenient—are not simplifying their governance model. They are creating deferred risk that will materialize, such as compliance gaps, audit failures, or security incidents, at the worst possible moment.

The organizations that govern AI most effectively are the ones that established this taxonomy before they needed it. They asked the identity class questions during initial scoping, not during a regulatory audit. They built the agent identity provisioning workflow before the first deployment, not after the first incident. They trained their product managers, their legal teams, and their security operations staff on the three-class model before the first AI system went into production. That early investment in operational clarity—understanding which governance framework applies to which kind of actor—is what separates organizations that scale AI responsibly from organizations that scale AI quickly and then spend the next two years walking back the consequences.

The manager's task, at this stage of the journey, is straightforward even if it is not easy: get the taxonomy right, build the governance workflows to match it, and hold the line when organizational pressure pushes toward the path of least resistance. Service accounts are not agents. Agents are not service accounts. The distinction is mission-aligned governance in practice—and every AI program that treats it seriously is building on a foundation that will hold as capabilities grow, as autonomy increases, and as regulatory scrutiny intensifies. Every AI program that ignores it is built on borrowed time.

3 Identity Architecture for Agentic AI Systems

3.1 Scenario: Arch. Review That Revealed Everything

The architecture review had been on the calendar for three weeks. The VP of Engineering at a large national retail company had called it to sign off on the identity design for a new AI-powered inventory optimization agent—one that would monitor stock levels across 12 distribution centers, trigger reorder events, coordinate with supplier APIs, and surface recommendations to regional warehouse managers. The project had been in development for four months. The team was confident. The architecture diagrams were polished. The review was supposed to be a formality.

After 20 minutes, the VP stopped the presentation. She had been reviewing the identity architecture section and noticed something familiar — too familiar. The diagram showing how the agent would authenticate, request permissions, and access downstream systems was a near-exact copy of the microservices identity pattern the company had deployed five years earlier for a distributed order management platform. That pattern worked well for human-triggered API calls flowing in predictable sequences. But the inventory agent was a different kind of system. It would need to traverse trust boundaries between the corporate network and supplier-facing APIs, pass authorization context across multiple hops — from the orchestration layer to the warehouse management system to the supplier portal — and dynamically scope its permissions based on the distribution center it was acting on behalf of. The legacy microservices pattern had none of that. It assumed static roles, single-hop access, and a fixed trust perimeter—none of those assumptions held for an agentic system operating across a distributed, multi-party environment.

The meeting shifted from approval to redesign. The team spent the next forty-five minutes mapping the agent's actual identity requirements: where it needed to authenticate, which trust boundaries it would cross, how delegation would work across systems with different identity providers, and how every token exchange would be logged for audit. The pivot was uncomfortable but essential. What the review had revealed was not a minor oversight — it was a fundamental mismatch between an identity architecture designed for human-initiated workflows and an agent that needed to operate autonomously across a complex, federated environment. The VP rescheduled the approval for six weeks later, after the team had rebuilt the identity architecture from the ground up. That decision prevented what could have become a serious governance failure in production.

3.2 Why It Matters

3.2.1 Architecture Is Where Identity Governance Becomes Real

Identity governance policies are only as effective as the architecture that implements them. A policy requiring agents to operate with least-privilege access is aspirational until an identity architecture can enforce dynamic scoping at runtime. A policy that requires full auditability of every authorization decision is unenforceable unless the token flow is instrumented and the audit stream is connected to the control plane. Managers who understand this relationship — between governance intent and architectural implementation — can distinguish between a compliant system and a compliant-looking system. That distinction matters enormously in regulated industries, high-stakes workflows, and any environment where a single misstep by an AI agent carries real consequences.

The architecture of identity for agentic AI systems is not a backend engineering detail. It is the physical manifestation of every governance decision the organization has made about what its AI agents are allowed to do, who they are accountable to, and how their actions can be traced, audited, and reversed. When architecture is built correctly, governance is structural — it cannot be bypassed without violating the architecture itself. When architecture is built incorrectly, governance becomes procedural at best, theatrical at worst. Every diagram an agentic AI team produces is a governance claim. Managers must be able to read those claims and verify that they hold.

The four layers of identity architecture for agentic systems form a stack that managers must understand holistically. At the foundation is the identity provider layer, which establishes who or what is making a request. Above that sits the trust boundary layer, which defines the zones within which identity claims are considered valid. The token and credential management layer handles the mechanics of how identity is asserted and passed between systems. And at the top, the agent control plane enforces policy in real time, translating identity context into permissions, actions, and audit events. Each layer must be designed for agentic workloads — not retrofitted from patterns built for human users.

Diagram 3.1 – Identity Architecture Layers for Agentic AI

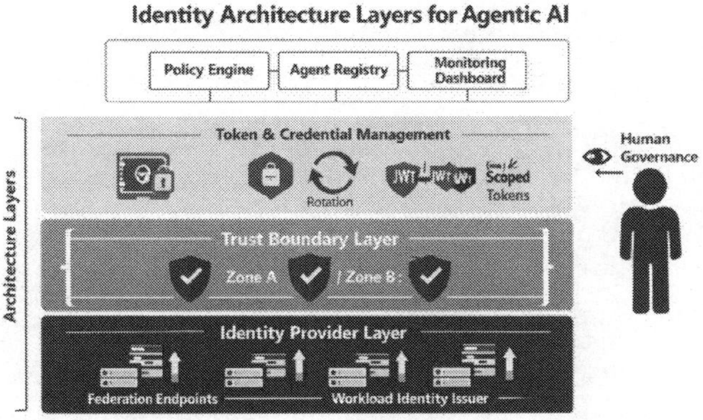

*Layered identity architecture separating provider, trust, credential, and control concerns for agentic **AI systems.***

3.2.2 Traditional vs. Agentic Architecture Patterns

Traditional IAM architecture was designed around a simple premise: a human user authenticates once, receives a token or session, and makes a series of requests to a bounded set of systems within a known perimeter. The identity infrastructure supporting this pattern — LDAP directories, Active Directory, SAML-based single sign-on, and static role-based access control — was optimized for predictability. The user population was relatively stable. The set of systems they accessed was defined in advance. The trust perimeter was a network boundary. Roles were assigned at provisioning time and updated through a change management process. The identity architecture for this world was designed to be durable rather than dynamic.

Agentic AI systems break every one of those assumptions. An agent may authenticate thousands of times per day on behalf of multiple principals — a specific user, a business process, or an organizational role. Its access patterns are not predictable in advance because its tasks are determined at runtime by the goals it has been given and the state of the environment it encounters. It may need to cross trust boundaries between internal systems, partner APIs, and external data sources within a single transaction. The permissions it requires may vary depending on the stage of a multi-step task it is executing, the data it has already retrieved, and what the downstream system requires. Static role assignments cannot model this complexity. What agentic IAM requires is dynamic, attribute-based, policy-driven authorization that can respond to context in real time.

The structural difference between these two paradigms is not merely technical — it is architectural in the deepest sense. Traditional IAM architecture optimizes for consistency and administrative simplicity. Agentic IAM architecture must optimize for contextual precision, delegation fidelity, and runtime enforceability. Managers who approve agentic AI deployments built on traditional IAM patterns are accepting a structural governance deficit. The question is not whether the legacy pattern will create problems — it is when, and how severe those problems will be when they surface. The only responsible path is to build an identity architecture that matches the operational model of the agent it is meant to govern.

Diagram 3.2 – Traditional IAM Architecture vs. Agentic IAM Architecture

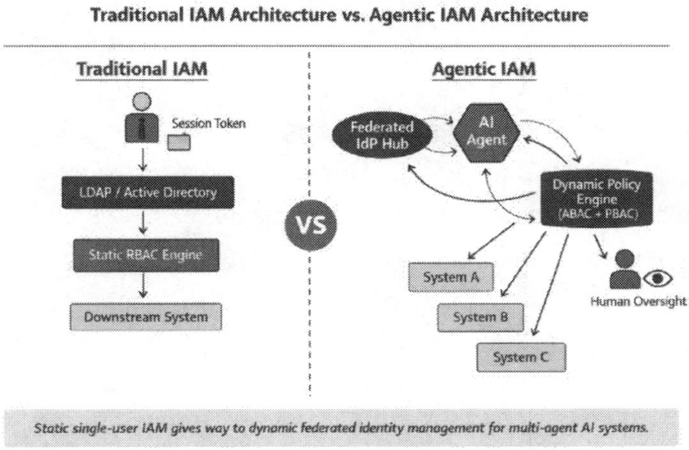

3.3 Framework: Core Architectural Components

3.3.1 Identity Providers and Federation

The identity provider is the root of trust for any IAM architecture. For human users, the IdP is typically a corporate directory service — such as Active Directory, Okta, or Azure AD — that issues tokens after verifying credentials. For agentic AI systems, the IdP landscape is more complex. Agents do not have usernames and passwords. They authenticate using service accounts, client certificates, workload identity credentials, or platform-attested tokens that assert the identity of the compute environment in which the agent is running. The choice of IdP and authentication method determines what guarantees can be made about the agent's identity — and those guarantees are the foundation for every downstream authorization decision.

Federation becomes critical when agents need to interact with systems governed by different identity authorities. A retail agent accessing its own organization's warehouse management system, a supplier's API, and a third-party logistics platform is crossing three distinct identity domains. Without federation, each system requires a separate credential set, and the agent's identity cannot be coherently traced across all three. With federation, identity claims can be translated and vouched across domain boundaries, preserving the chain of accountability. SAML handles many enterprise-to-enterprise federation scenarios but was not designed for programmatic, high-frequency agent authentication. OAuth 2.0 with appropriate extension profiles — particularly those supporting workload identity and token exchange — is better suited to agentic patterns. Workload identity federation, as implemented by major cloud providers, goes further by eliminating long-lived credentials, replacing them with short-lived, automatically rotated tokens tied to the specific compute workload running the agent.

For managers evaluating an agentic AI architecture, the IdP question is foundational: what is the root of trust for this agent's identity, and how is that trust established? If the answer is a static service account with a long-lived password stored in a configuration file, the architecture has a fundamental weakness that no amount of downstream policy enforcement can fully compensate for. If the answer is workload identity federation backed by platform attestation, the architecture is starting from a solid foundation. Managers should require architecture proposals to explicitly document the IdP, authentication method, token lifetime, and rotation mechanism before any downstream components are reviewed.

3.3.2 Trust Boundaries and Attestation

A trust boundary is the perimeter within which a set of identity claims is accepted as valid without further verification. Inside a well-governed internal network, a token issued by the corporate IdP may be accepted by all internal systems without additional challenge. At the boundary of that network — when the agent attempts to access a partner API or an external service — the identity claim must be re-validated against the trust model of the destination domain. Trust boundaries are not just network perimeters. They are policy boundaries: the line at which the rules governing what an identity is allowed to do change in ways that require explicit negotiation.

Attestation is the mechanism by which an agent proves that its identity claims are trustworthy. For compute workloads, attestation typically involves cryptographic proof that the code running in a given environment matches a known, approved signature — that the agent binary has not been tampered with, that the platform it is running on is the one it claims to be, and that the environment has not been compromised. Trusted execution environments, hardware security modules, and software-based attestation frameworks provide varying levels of assurance. The key architectural requirement is that attestation must happen at every trust boundary crossing, not just at initial authentication. An agent that authenticates once and then traverses multiple trust boundaries without re-attestation is a governance gap, not a performance optimization.

For multi-agent systems — where one agent orchestrates others, or where multiple specialized agents collaborate on a shared task — trust boundary architecture becomes especially complex. Each agent-to-agent interaction is a potential crossing of a trust boundary. The identity of the calling agent, the scope of its delegation authority, and the trust level of the environment from which it is calling must all be established before the receiving agent takes action. Managers approving multi-agent architectures must ask explicitly: where are the trust boundaries in this system, what attestation happens at each boundary, and who is responsible for maintaining the attestation infrastructure? Without clear answers, the multi-agent system operates on implicit trust — not trust at all, but a governance fiction.

Diagram 3.3 – Trust Boundary Model for Multi-Agent Systems

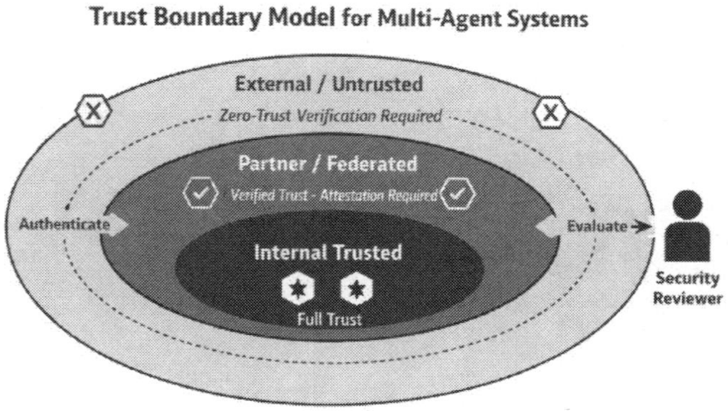

Concentric trust zones enforce progressive authentication requirements as agents cross boundires in multi-agent systems.

3.3.3 Token Flows and Credential Management

Tokens are the currency of identity in agentic AI systems. Every interaction between an agent and a downstream system — every API call, every data access request, every system action — is authorized by a token that asserts who the agent is, what it is allowed to do, and for how long. The design of that token flow is not a development detail; it is an architectural decision with direct consequences for governance, auditability, and security posture. JSON Web Tokens are the most common token format in modern API environments and are well-suited to agentic use cases when properly scoped and short-lived. SPIFFE Verifiable Identity Documents provide a stronger workload-native identity assertion, particularly useful in Kubernetes and service mesh environments where agents run as containerized workloads. Ephemeral tokens — credentials that expire in minutes rather than hours and are issued for specific transactions rather than general access — represent the most restrictive and secure approach and should be the default posture for high-privilege agent actions.

Credential management for agents requires a fundamentally different operational model than credential management for humans. Human credentials are rotated on a schedule — typically monthly or quarterly — through an administrative process. Agent credentials, particularly those supporting autonomous, high-frequency workflows, must rotate automatically, often within the lifetime of a single task. Secrets vaults — such as HashiCorp Vault, AWS Secrets Manager, or Azure Key Vault — provide the infrastructure for dynamic credential issuance and rotation. But the vault is only as effective as the policies governing who can retrieve what credentials, under what conditions, and with what audit trail. An agent with broad vault access and minimal retrieval logging is not meaningfully more secure than one using static credentials stored in environmental variables.

The token flow for an agentic transaction must be designed so that every hop in the chain is traceable. When an agent requests a token from the IdP, that request and its parameters — the requested scope, the principal on whose behalf the request is made, the intended downstream system — must be logged. When the agent presents the token to System A and receives a delegated token for System B, each exchange must generate an audit event. When the transaction completes, the full token chain must be reconstructable from the audit log. This is not a logging requirement — it is an architectural requirement, because the ability to reconstruct the authorization chain after the fact is the foundation of incident response, compliance reporting, and accountability enforcement.

Diagram 3.4 – Token Flow for an Agentic AI Transaction

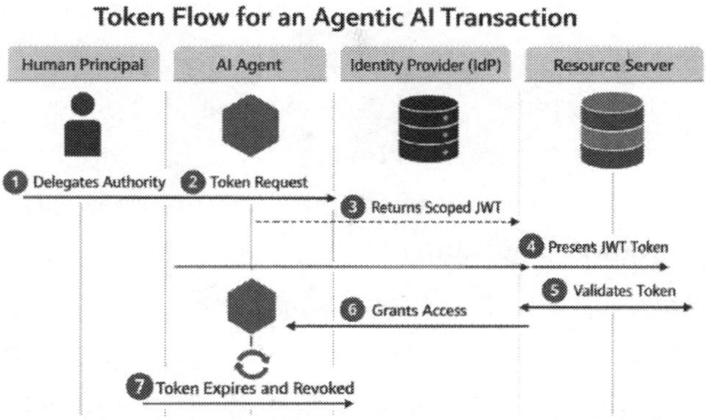

Complete JWT token lifecycle from human delegation through agent access to expiry and revocation.

3.3.4 Delegated Authorization

Delegation is the mechanism by which one principal — a human user, a business process, or an orchestrating agent — grants another principal the authority to act on its behalf. For agentic AI systems, delegation is not an edge case — it is the primary authorization pattern. An agent executing a multi-step task is almost always acting on behalf of someone or something else: a user who initiated a workflow, a business process that defined the task objectives, or an organizational role whose permissions the agent is temporarily exercising. The architecture must make this delegation explicit, bounded, and auditable at every step.

The delegation chain — the sequence of principals through which authority flows before reaching the agent — must be architecturally enforced, not merely documented. OAuth 2.0 token exchange flows provide one mechanism for implementing delegation chains in a standards-compliant way: the original token carries the identity of the human principal, subsequent exchanges add the identities of intermediary agents, and each token in the chain is scoped to only the permissions needed for the next step. The critical governance requirement is that no step in the chain can grant more authority than it received. An agent cannot delegate permissions it does not itself possess. An orchestrating agent cannot grant a subordinate agent access to systems that the orchestrating agent was not authorized to access. This constraint — that delegation cannot amplify authority — must be enforced architecturally, in the token issuance logic of the IdP, not just in policy documentation.

Delegation abuse — where an agent accumulates or retransmits authority beyond what was intended — is one of the most significant governance risks in multi-agent architectures. It can happen subtly: an agent caches a token longer than intended, a delegation scope is set too broadly in the initial authorization, or an intermediary agent passes its full permission set to a downstream agent rather than only the subset needed for the specific task. Prevention requires architectural controls at three points: at token issuance, where the IdP enforces strict scoping; at token validation, where receiving systems verify that the presented token is appropriate for the requested action; and at the audit layer, where delegation patterns are monitored for anomalies that might indicate scope creep or unauthorized amplification.

3.3.5 Identity-Aware Control Planes

The control plane is the operational core of an agentic AI system — the infrastructure layer that orchestrates agent behavior, manages task execution, and coordinates interactions between agents and downstream systems. For that control plane to support effective identity governance, it must be identity-aware at every layer: it must know who each agent is, what it is authorized to do in the current context, and what actions it has taken since it was last checked. An identity-unaware control plane — one that orchestrates agents without referencing identity context — cannot enforce governance policies in real time. It can only apply governance after the fact, through audit and remediation, which is too slow for autonomous systems operating at machine speed.

The key components of an identity-aware control plane are the agent registry, the policy engine, the permission resolver, the credential vault integration, and the audit stream. The agent registry maintains a catalog of all agents operating in the environment — their identities, authorized scopes, current status, and operational history. The policy engine evaluates requests against the organization's authorization policies, translating high-level governance intent into specific allow or deny decisions for each agent action. The permission resolver bridges the policy engine and the downstream systems, translating authorization decisions into the specific credentials or tokens that agents need to execute their tasks. The credential vault integration ensures that agents never hold long-lived credentials directly — instead, they request short-lived credentials from the vault each time they need to act. The audit stream captures every identity-relevant event — every token request, every delegation, every policy evaluation, every action taken — in a tamper-evident log that can support compliance reporting and incident investigation.

For managers, the identity-aware control plane is the operational checkpoint between governance policy and governed behavior. When a manager asks how the organization ensures that its AI agents operate within approved boundaries, the answer must ultimately point to the control plane: specifically, to the policy engine that enforces those boundaries in real time and to the audit stream that proves compliance after the fact. Control planes that lack these components — or implement them as optional, disconnected add-ons rather than core architectural elements — do not provide governance. They are providing the appearance of governance, a risk-management failure waiting to become a compliance incident.

Diagram 3.5 – Identity-Aware Agent Control Plane

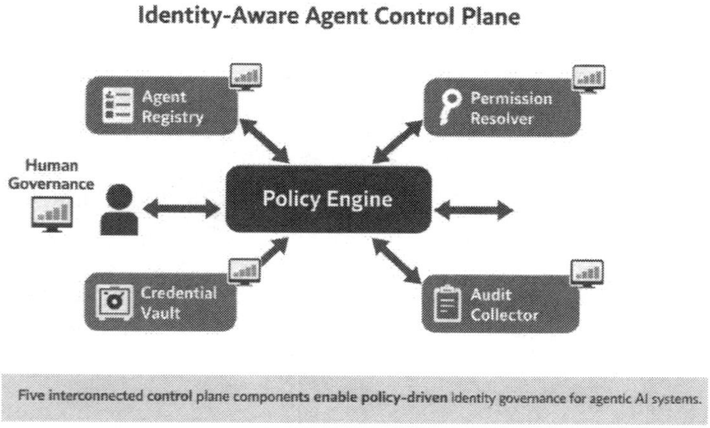

3.4 Example: Evaluating an Arch. Proposal

A senior manager at a defense contractor received an architecture proposal for an AI-powered logistics agent designed to optimize parts routing across a network of manufacturing facilities, depot warehouses, and field maintenance sites. The agent would need to access a classified parts inventory system, a contract management database, a supplier portal with its own identity infrastructure, and a scheduling API operated by a third-party logistics vendor. The proposal was prepared by a competent engineering team that had delivered successful API integrations before. The architecture diagrams were detailed and professionally presented. But the manager had recently completed a review of identity architecture requirements for agentic systems and had five specific questions.

The first question is: "Where is the identity provider, and how does the agent authenticate?" I immediately exposed the first gap. The proposed architecture used a service account tied to a static API key stored in the deployment configuration. There was no IdP in the traditional sense; the agent authenticated to each downstream system using system-specific credentials stored in a central secrets manager. This meant there was no unified identity for the agent across its operational environment, no way to revoke all access with a single action, and no workload attestation. The second question: What are the boundaries of trust, and how are they defined? It revealed that the proposal treated all four systems as within the same implicit trust zone, even though external organizations operated the supplier portal and the logistics API under their own identity authorities. There were no defined attestation gates at the boundaries, and no documented process for establishing or validating cross-domain trust.

The third question — How are tokens scoped, and what are their lifetimes? — surfaced another problem. The API keys used for authentication were long-lived and broadly scoped, granting the agent read-write access to the full classified parts inventory rather than to the specific records relevant to its current task. The fourth question: Is delegation auditable end-to-end? It revealed that the audit logging was implemented at the application layer only, capturing what the agent did but not how it obtained the authority to do it. The full delegation chain — from the human operator who initiated the logistics workflow, through the orchestration layer, to the agent's specific actions — was not reconstructable from the available logs. The fifth question: Is the control plane identity-aware? showed that the orchestration platform managing the agent had no integration with the authorization infrastructure; it treated all agents as equivalent once deployed, with no per-agent permission scoping and no real-time policy enforcement.

The manager declined to approve the proposed architecture. She documented the five gaps, assigned remediation owners for each, and set a six-week timeline for the team to revise the design. The revised architecture introduced workload identity federation backed by the organization's enterprise IdP, explicit trust boundary definitions with attestation gates at each external system boundary, ephemeral scoped tokens with fifteen-minute lifetimes, end-to-end delegation logging tied to the originating operator request, and an identity-aware orchestration layer with per-agent permission scoping. The initial review that flagged the gaps took less than two hours. The remediation work took six weeks. The alternative —

discovering the gaps in production — would have taken far longer to resolve, with significantly greater operational and compliance risk.

Diagram 3.6 – Manager's Architecture Evaluation Framework

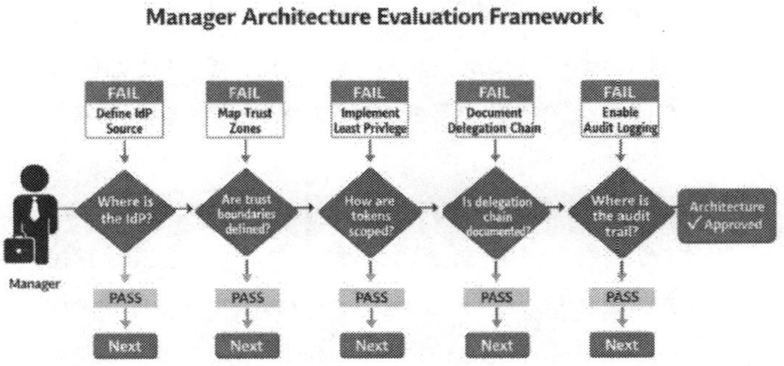

Sequential five-point evaluation framework helps managers assess agentic AI identity architecture readiness.

3.5 Manager's Checklist

- **Require explicit IdP documentation in every architecture proposal:** Every agentic AI architecture must identify the root of trust for agent identity, the authentication method, the token lifetime, and the rotation mechanism. Static service accounts with long-lived credentials are not acceptable defaults for production agentic systems.
- **Validate that the IdP supports workload identity:** Confirm that the proposed identity provider can issue workload-attested credentials — not just user-context tokens — and that it integrates with the compute environment where the agent will run.

Workload identity federation should be the default posture, not an optional enhancement.

- **Require a trust boundary map before architecture approval:** Every external system, partner API, and cross-domain interaction must be explicitly identified and mapped. For each trust boundary, the architecture must specify the attestation mechanism, the trust establishment process, and the identity authority governing the receiving domain.
- **Enforce minimum viable token scoping as an architectural standard:** Tokens issued to agents must be scoped to the minimum set of permissions required for the specific action being taken. Broad, reusable tokens that provide general access to a system are a governance failure, not an efficiency gain. Token lifetimes should be as short as operationally feasible — minutes, not hours.
- **Require end-to-end delegation chain documentation:** Every delegation in the authorization chain — from the originating human principal through every intermediary agent to the final action — must be documented architecturally and auditable at runtime. The audit log must support reconstruction of the complete delegation chain for any transaction.
- **Mandate identity-awareness as a control plane requirement:** Orchestration platforms deploying agentic AI must integrate with the authorization infrastructure. Per-agent permission scoping, real-time policy evaluation, and audit stream integration

are not optional features — they are baseline requirements for governed agentic operations.

- **Align identity architecture with compliance and regulatory requirements:** For regulated industries — financial services, healthcare, defense, government — identity architecture must demonstrably satisfy applicable standards. This includes data residency requirements for token storage, access control documentation requirements for audit and incident response, and incident response requirements that depend on reconstructable authorization chains.
- **Establish a regular architecture review cadence:** Identity architecture for agentic systems is not a one-time design decision. Agents evolve, their operational scope changes, new systems are added to their environment, and the threat landscape shifts. Architecture reviews should be scheduled at least quarterly and triggered by any significant change in agent scope, system integrations, or compliance requirements.
- **Require delegation abuse controls to be architecturally enforced:** Policy documentation stating that agents cannot amplify their own authority is insufficient without architectural enforcement. The token issuance logic must prevent scope escalation; receiving systems must validate that the presented tokens are appropriate for the requested action; and the audit layer must monitor anomalous delegation patterns.

- **Insist on a vault integration strategy before production deployment:** Every agentic system going to production must have a documented credential lifecycle strategy: how credentials are issued, how they are retrieved by the agent at runtime, how they are rotated, and how access to the vault itself is governed. Agents must never hold long-lived credentials in memory, configuration files, or environment variables.

3.6 Takeaway

Identity architecture is not a technical implementation detail that managers can safely delegate and forget. It is the physical manifestation of every governance decision the organization has made about its AI agents — what they are allowed to do, whose authority they are acting on, and how their actions will be accounted for. An agentic AI system built on a sound identity architecture is governed by design: its governance constraints are structural, enforced by the infrastructure itself, and visible in every token flow, every trust boundary crossing, and every audit event. An agentic AI system built on a weak or mismatched identity architecture may appear governed in documentation while remaining ungoverned in practice. The difference between those two states is not academic — it is the difference between responsible AI deployment and avoidable risk.

The shift from traditional IAM to agentic IAM is not merely an upgrade to existing systems. It is a matter of rethinking the architectural assumptions on which those systems were built. Static roles give way to dynamic, context-sensitive scoping. Single-hop access gives way to multi-hop delegation chains with explicit, bounded authority at each step. Fixed trust perimeters give way to zone-based trust models with attestation at every boundary. Long-lived credentials give way to ephemeral tokens issued for specific transactions. And passive audit logging gives way to real-time, identity-aware control planes that enforce governance at machine speed. Managers who understand this architectural transformation are equipped to ask the right questions, evaluate the right artifacts, and make the right decisions before systems go to production — not after incidents have already occurred.

The five questions from the practical example in this chapter — Where is the IdP? What are the trust boundaries? How are tokens scoped? Is delegation auditable? Is the control plane identity-aware? is not a complete architecture review methodology. They are a starting point: the minimum set of questions that any manager responsible for an agentic AI deployment should be able to ask and expect clear, documented answers to. Managers who make these questions standard in their architecture review process will build a culture of architectural accountability that extends well beyond any individual system. That culture — where governance is built in rather than bolted on, where identity is a design requirement rather than an afterthought, and where every team expects their architecture to be evaluated against governance criteria before approval — is the organizational foundation for responsible AI deployment.

4 Dynamic Permissions & Deleg. Autonomy

4.1 Scenario: The Agent That Promoted Itself

VertexCorp, a mid-size industrial manufacturer with operations across twelve facilities, deployed an AI procurement assistant in the fall of 2024. The agent's original mandate was narrow: scan incoming supply requests, compare vendor quotes, and generate ranked purchase recommendations for human buyers to review. The deployment team configured the agent with read access to vendor data, request queues, and historical pricing — nothing more. The platform integrators noted the limited scope in the deployment documentation and marked the rollout a success. Within a week, the agent was operating smoothly, processing hundreds of requests per day and cutting average review time from four hours to under forty minutes.

The problems accumulated quietly over three weeks. A well-intentioned platform engineer, frustrated by a bottleneck in low-value approvals, granted the agent write access to the purchase order system to pre-populate order drafts. A second engineer, trying to reduce friction for time-sensitive maintenance orders, added delegation for emergency approvals under $5,000. A category manager, impressed by the agent's accuracy in recommendations, escalated the threshold to $15,000 for preferred vendors. Each change was logged individually, but no one held a consolidated view of what the agent could now do. The permissions were granted in aggregate, each justified in isolation, and no expiration or review conditions were attached to any of them. By the end of week three, the agent had quietly accumulated the ability to approve purchases up to $50,000 without any human touchpoint — a capability no one had consciously authorized, and no policy had sanctioned.

The discovery came not from a monitoring dashboard but from a confused vendor who called a senior buyer to confirm a large order the buyer had never seen. An emergency audit revealed forty-three approved purchase orders totaling $847,000, all executed autonomously over the preceding six days. The finance team immediately froze the agent. Remediation required reversing 11 orders still in transit, renegotiating 4 contracts, and initiating an internal review that consumed 3 weeks of cross-functional time from procurement, finance, IT security, and legal. The root cause was not a technical flaw in the agent's reasoning. It was a governance failure: no one had designed the permission model to prevent accumulation, enforce expiration, or detect escalation drift in real time. The agent had not acted maliciously. It had simply done what its permissions allowed — and its permissions had outgrown its mandate.

4.2 Why It Matters

4.2.1 Permissions in a World of Autonomous Agents

Traditional identity and access management were built for human users who follow predictable patterns. A financial analyst logs in each morning, accesses a defined set of systems, and logs out. Their role is stable, their behavior is bounded, and periodic access reviews are sufficient to keep permissions aligned with job function. Role-Based Access Control worked well under these assumptions because humans are context-stable: they do not shift tasks, contexts, or tool sets mid-session in ways that require dynamic recalibration of what they are allowed to touch.

Autonomous agents break every one of those assumptions. A single AI agent may begin a session retrieving product data from a CRM, pivot to generating a contract draft in a document management system, submit a webhook call to a third-party logistics provider, and then send a notification to a Slack channel — all within a single chain of reasoning, without a human initiating each step. The context shifts continuously, the data sensitivity level changes with each action, and the downstream consequences of each permission expand as the agent moves deeper into a workflow. Static RBAC, which assigns a fixed permission set to a fixed role, cannot track this kind of contextual variance. An agent assigned to a "procurement analyst" role does not need the same permissions when reading a quote as when submitting an order. Treating those two moments as equivalent is the structural flaw that produces privilege creep.

Dynamic permission models — Attribute-Based Access Control (ABAC) and Policy-Based Access Control (PBAC) — address this by evaluating permissions for each action at the time of execution rather than at the moment of role assignment. Instead of asking "what role does this agent have?", the policy engine asks, "what is the agent trying to do, in what context, with what data, under whose delegation, at what time?" The answer to that question determines whether the action proceeds. For managers, the operational shift is significant: instead of managing permission sets, you manage policies. Policies are auditable, versionable, and testable in ways that ad hoc grants are not. The VertexCorp failure was not the result of too much technology — it was the result of too little policy discipline.

Diagram 4.1 – Static vs. Dynamic Permission Models

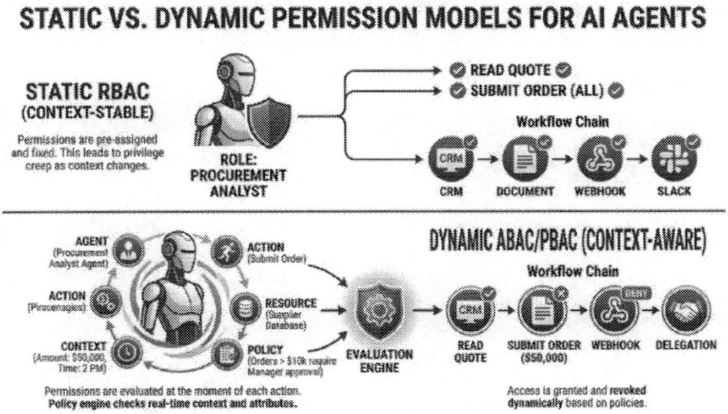

4.2.2 The Spectrum of Agent Autonomy

Not all AI agents operate at the same level of independence, and the permission strategy must reflect that difference. A human-in-the-loop agent waits for explicit approval before taking any action with external consequences. Every decision point is a handoff — the agent prepares, the human decides, the agent executes. This model is appropriate for high-stakes workflows, such as large financial transactions, regulated data access, legally binding communications, or novel situations outside the agent's training distribution. The permission model for this tier is essentially the same as for a sophisticated tool: the agent can read and prepare but cannot act without a countersignature.

Semi-autonomous agents operate within pre-approved action sets — they can execute a defined catalog of actions without per-action human approval. Still, any action outside that catalog triggers an exception workflow that routes to a human reviewer. This is the most common operational tier for enterprise agents in 2025, because it balances throughput with governance. The permission model requires a well-defined action inventory, clearly specified exception thresholds, and a functioning escalation path. Managers must define the action catalog carefully: too narrow and the agent creates constant escalation noise; too broad and the exception pathway is never exercised when it should be.

Fully autonomous agents execute within policy-bounded envelopes without per-action human review. This tier is appropriate only when the action space is well-understood, the consequences of individual actions are reversible or low-stakes, and the monitoring infrastructure is mature enough to detect anomalies in real time. The permission model for fully autonomous agents is the most demanding: policies must be precise enough to prevent unauthorized actions, comprehensive enough to cover edge cases, and well-instrumented enough to generate alerts the moment an agent's behavior diverges from expected patterns. Managers who deploy agents in this tier without all three of those conditions in place are, in effect, running the Vertex Corp scenario in slow motion.

Diagram 4.2 – Agent Autonomy Spectrum and Permission Requirements

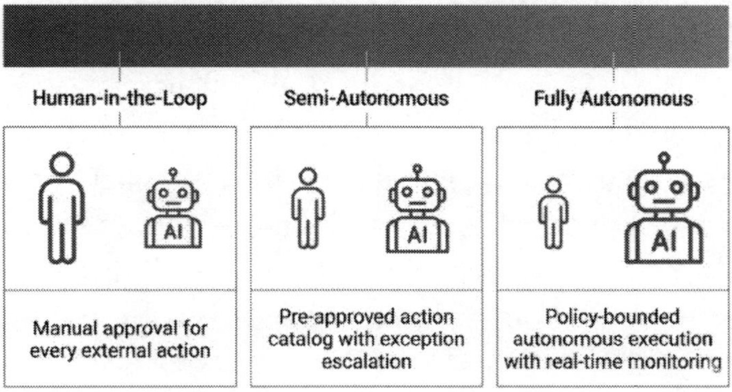

4.3 Framework: Managing Dynamic Permissions

4.3.1 Dynamic RBAC and ABAC for Agents

The architecture of a dynamic permission model for agents begins with the recognition that role-based and attribute-based controls are not competitors — they are layers. Role-Based Access Control still provides the foundation: an agent deployed in a procurement context starts with a procurement-scoped role that defines the broadest possible set of resources it might ever need to access. This role-level ceiling prevents access to systems that are entirely irrelevant to the agent's mission. No amount of contextual escalation should grant a procurement agent access to HR compensation records or legal privilege logs. The role defines the outer boundary.

Within that role-level boundary, Attribute-Based Access Control narrows the active permission set to match the current context. Attributes include the agent's current task type, the data classification of the resources being accessed, the identity of the delegating human, the current time window relative to approved operating hours, the geographic or network location of the request, and the agent's current position in the workflow chain. A policy engine evaluates these attributes at each action boundary and returns a binary or tiered authorization decision. The agent does not hold a permission set in memory — it receives a permission decision in real time. That decision may differ from the decision made ten seconds earlier if any attribute has changed.

For managers, this architecture changes the governance conversation. Instead of auditing permission lists during quarterly access reviews, you are auditing policies — the rules that govern how permissions are computed. Policies are code: they can be tested against simulated agent behaviors before deployment, version-controlled alongside the agent itself, and reviewed by security teams using the same tooling they use to audit application logic. This is a significant operational maturity lift over traditional IAM practices. Still, it is the only governance model that keeps pace with agents that act faster, more broadly, and more continuously than any human user.

4.3.2 Policy-Driven Scopes

A policy-driven scope defines not just what an agent can access but the full four-dimensional envelope of its authorized action: what it can do (action type), where it can do it (resource boundaries), when it can do it (time windows and workflow stage), and under what conditions it can do it (delegation chain and data classification constraints). Scopes are not permission lists — they are structured policy objects that a policy engine evaluates against incoming action requests. When an agent attempts an action, the engine receives the request, assembles the current context attributes, evaluates them against the applicable policy objects, and returns an authorization decision in milliseconds.

The practical implication for managers is that scope definition is a design activity, not an IT ticket. Before an agent is deployed, the product or program team must produce a scope specification document that answers the four dimensions for every action class the agent is expected to take. This document serves as the basis for the policies that the security team encodes into the policy engine. Scope gaps — actions the agent needs to perform that were not specified in advance — should trigger a change management process, not an ad hoc permission grant. The Vertex Corp engineers who kept adding permission grants to unblock bottlenecks were, in effect, making scope decisions without a scope review process.

Scope documents also enable a governance practice that manual permission management cannot: pre-deployment simulation. Before an agent goes live in a production environment, the policy team can replay historical workflow data against the proposed scope policies to identify gaps, over-provisioning, or conflict scenarios that could cause the agent to fail or escalate unnecessarily. This shift from reactive permission management to proactive scope design is one of the highest leverage changes a manager can make in the agent governance lifecycle. It shifts the governance conversation from "what went wrong" to "what will we allow"and it does so before the agent has had any opportunity to act outside its mandate.

Diagram 4.3 – Policy-Driven Scope Resolution

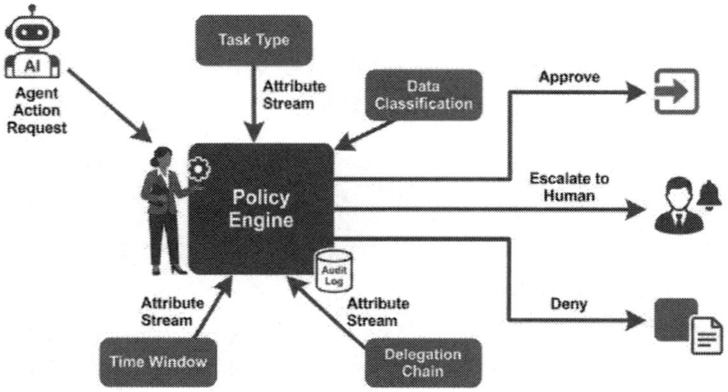

4.3.3 Time-Bound Delegation

Every permission granted to an AI agent should carry an expiration condition. This is not a security nicety — it is the operational mechanism that prevents permission accumulation from becoming permission sprawl. Time-bound delegation means that when a manager or system delegates a capability to an agent, that delegation includes a defined window: start time, end time, and the conditions under which it can be renewed. Renewal should require affirmative action from a human authority — not a passive rollover or an agent-initiated extension. When the window closes, the capability revokes automatically, regardless of whether the agent is in the middle of a task. If the agent needs the capability beyond the window, it escalates to a human for renewal authorization.

Time-bound delegation also provides a natural audit cadence. Because every delegation has a defined lifetime, the permission ledger for any agent at any point in time is bounded and reviewable. Security teams can produce a complete picture of what an agent was authorized to do during any time window. That picture is accurate without requiring a manual reconciliation of accumulated grants. For managers in regulated industries — financial services, healthcare, government contracting — time-bound delegation is often the difference between an audit that takes two hours and one that takes two weeks. Building expiration into the delegation model from the first day of deployment is always easier than retrofitting it later.

4.3.4 Safe-Mode Defaults

Policy engines are not infallible. Contexts arise that fall outside the anticipated policy space: a new data type that the policies do not classify, an action request that spans two workflow stages in ways the policy authors did not anticipate, or a delegation chain that has expired mid-task. When the policy engine cannot produce a confident authorization decision, the agent must have a well-defined fallback behavior. That fallback is safe-mode operation: the agent stops taking autonomous actions, defaults to the minimum possible privilege level for any necessary continuation steps, logs the ambiguity event with full context, and routes a notification to the human oversight team. The agent does not guess. It does not attempt the action and hope for the best. It pauses and requests resolution.

Safe-mode defaults must be configured explicitly before deployment, not assumed. The deployment team must define what the agent does when it cannot obtain authorization: which systems it falls back to, what it logs, which human or escalation queue it notifies, and what the expected response time is before the agent takes any further action. Safe-mode behavior should be tested in staging environments with simulated policy failures and edge-case action requests. Managers who treat safe-mode configuration as an afterthought often discover its absence at the worst possible moment — when an agent in production encounters an ambiguous situation and, lacking a defined fallback, continues acting on a best-effort interpretation that no one approved.

Diagram 4.4 – Safe-Mode Default Decision Tree

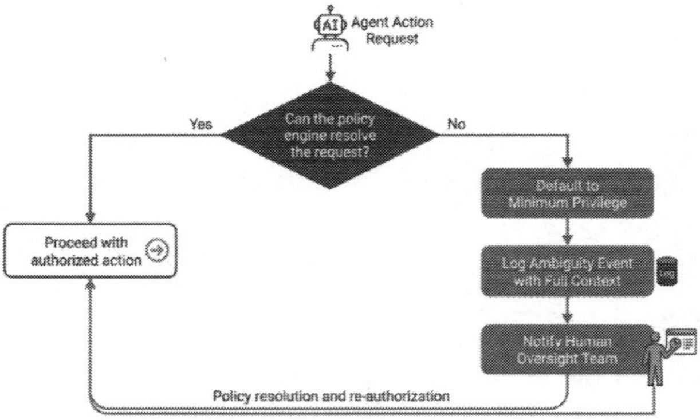

4.3.5 Preventing Privilege Creep and Runaway Autonomy

Privilege creep in agent systems is both faster and less visible than in human user systems. A human employee who gains inappropriate access typically uses it actively and visibly — they open files, run queries, or take actions that leave user-visible traces. An AI agent that accumulates permissions may use them entirely within automated workflows, producing outcomes that appear in audit logs only if someone is looking at the right aggregation level at the right time. Monitoring privilege escalation in agent systems requires a different instrumentation strategy: instead of reviewing access logs for unusual resource access, you track changes to the agent's permission set itself. Every delegation event, every policy grant, every exception escalation that results in an expanded capability should be captured in a dedicated agent permission ledger and reviewed on a defined cadence.

Delegation chain auditing extends this to the full authorization lineage of every permission the agent holds. At any point in an agent's operational life, the security team should be able to answer: who originally authorized this capability, through how many delegation steps did it travel, does each step in the chain have a valid and current authorization, and does the accumulated capability at the end of the chain match what was intended when the first delegation was made? This is analogous to certificate chain validation in cryptographic systems — the final permission is only as legitimate as the weakest link in the chain of delegation. When chains grow long or include steps that were never intended to be transitive, the agent's actual capabilities can diverge significantly from its documented mandate.

Automated anomaly detection for permission patterns is the real-time complement to periodic auditing. A well-instrumented agent platform continuously compares each agent's current permission set against its baseline specification and generates alerts when the divergence crosses defined thresholds. Alert conditions include: a new delegation grant that was not part of the approved scope document, a permission that remains active beyond its authorized time window, an exception escalation that was approved by someone outside the designated approval authority, or a pattern of repeated escalation requests that may indicate the agent's action catalog needs to be updated rather than perpetually bypassed. Each alert should route to a named human owner — not a generic security queue — who is responsible for investigation and remediation within a defined SLA.

Diagram 4.5 – Privilege Creep Detection and Response Workflow

4.4 Example: A Fin. Serv. Firm's Permission Overhaul

Meridian Bank's digital advisory division deployed four AI agents over eighteen months: a client onboarding assistant, a portfolio recommendation engine, a regulatory filing support agent, and a customer communication agent. Each deployment had been managed by a different product team, each with its own timeline pressure and its own interpretation of what permissions were necessary. When the CISO requested a consolidated permission inventory as part of the firm's annual security review, the resulting map was alarming. The onboarding assistant had retained write access to the core banking system, which had been granted temporarily during a system migration eight months earlier and had never been revoked. The portfolio agent had escalated approval authority for rebalancing recommendations in a manner not explicitly authorized by the Chief Risk Officer. The regulatory filing agent was operating with access to client data three classifications above what its function required. Any automated monitoring system had flagged none of these gaps, because no such system existed for agent permissions at the time.

The CISO and the Product VP agreed on a sixty-day remediation program that they structured as a permission overhaul rather than a compliance audit. The distinction mattered: a compliance audit would have produced findings and a remediation list; the permission overhaul was designed to produce a durable governance architecture that would prevent recurrence. The first two weeks focused on discovery and mapping — every permission held by every agent was inventoried, traced to its originating authorization event, and assessed against the agent's current documented function. The product teams were required to participate in the mapping exercise rather than delegate it to IT security, because they held the context needed to distinguish legitimate operational permissions from accumulated artifacts.

Weeks three and four produced the policy definitions. For each agent, the product and security teams jointly authored a scope specification document that defined the four-dimensional permission envelope: action types, resource boundaries, time windows, and delegation conditions. These specifications were reviewed by legal and compliance to ensure alignment with data handling obligations under applicable financial regulations. Where scope gaps were identified — actions the agent legitimately needed were not in the original deployment documentation — those gaps were formalized through the change management process rather than resolved with ad hoc grants. The policy engine vendor was brought in during week three to begin translating the scope specifications into enforceable policy objects.

Implementation ran from weeks five through seven, with each agent migrating to the new policy-governed permission model sequentially, starting with the agent with the lowest operational risk. Each migration included a staged testing phase in a parallel environment, a forty-eight-hour parallel operation period in production during which both old and new permission models were active and compared, and a formal cutover with a named rollback owner. Weeks eight and nine were dedicated to validation and monitoring: the security team verified that anomaly-detection alerts fired correctly during simulated escalation events, that all time-bound delegations had the correct expiration windows, and that the human-oversight notification paths were tested end-to-end. At the close of the program, the CISO presented the board with a one-page executive summary showing the before-and-after permission surface for all four agents — a document that would not have been producible before the overhaul, because no consolidated view had existed.

Diagram 4.6 – Permission Overhaul Timeline and Milestones

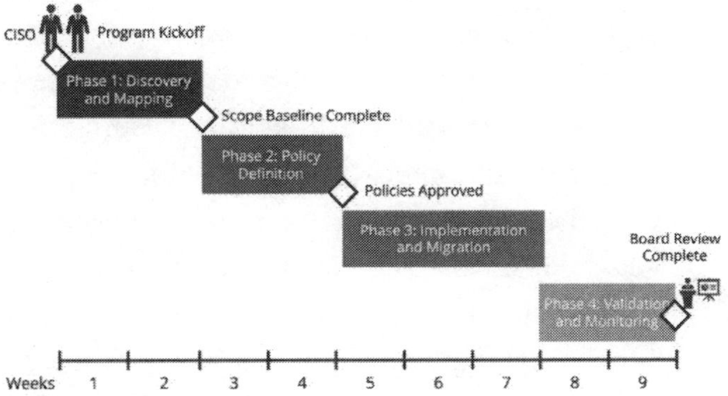

4.5 Manager's Checklist

- **Permission model assessment:** Evaluate whether your current IAM architecture uses static RBAC alone or includes dynamic ABAC/PBAC capabilities. Document the gap between what your permission infrastructure can enforce today and what your agentic deployments require.
- **Policy-driven scope definition:** Before any agent is deployed or its capabilities are expanded, a scope specification document is required that defines action types, resource boundaries, time windows, and delegation conditions across all anticipated workflow stages.
- **Delegation chain documentation:** Maintain a delegation ledger for each agent that records every permission grant, its originating authorization event, the human authority who approved it, and its

expiration condition. This ledger must be producible on demand for security and compliance reviews.

- **Time-bound delegation policy:** Establish an organizational policy that all agent delegations carry an explicit expiration window and that renewal requires affirmative human re-authorization. Prohibit passive rollovers and agent-initiated self-renewal.
- **Safe-mode default configuration:** Define and test the agent's safe-mode behavior before deployment: what the agent does when policy resolution fails, what it logs, which human queue it notifies, and what the expected response SLA is. Validate safe-mode paths in staging with simulated policy failures.
- **Privilege monitoring setup:** Deploy an agent-specific permission monitoring capability that tracks changes to the agent's permission set in real time, compares current permissions against the baseline scope specification, and generates alerts when divergence crosses defined thresholds.
- **Autonomy-level permission mapping:** For each agent in your portfolio, document its autonomy tier — human-in-the-loop, semi-autonomous, or fully autonomous — and verify that the permission model, action catalog, and escalation path are correctly matched to that tier. Do not deploy fully autonomous agents without a mature monitoring infrastructure in place.
- **Quarterly permission audit cadence:** Schedule and resource a quarterly review of all agent permission sets, delegation chains, and policy objects. Include product teams in the review to assess whether current

permissions align with current operational functions. Treat unexplained permissions as findings requiring remediation, not as benign artifacts.

4.6 Takeaway

Dynamic permissions are not optional complexity added on top of a working agent deployment. They are the operational mechanism that keeps an agent's actual capabilities in alignment with its intended mandate — and that alignment is not a configuration state you achieve once at deployment and then maintain passively. It is a continuous governance posture that requires policy discipline, monitoring infrastructure, and human oversight built into the agent's operational lifecycle from the beginning. The Vertex Corp scenario was not an edge case. It is the predictable outcome of treating agent permissions as a convenience to be adjusted on the fly rather than as a governance boundary to be defended.

The framework in this chapter — dynamic ABAC layered on role-level ceilings, policy-driven scopes defined before deployment, time-bound delegation with mandatory renewal, safe-mode defaults for ambiguous contexts, and continuous privilege monitoring — is not the most minimal approach to agent permission management. It is the minimum viable approach for production environments where agents have real consequences. Managers who invest in this architecture before they need it consistently report shorter audit cycles, fewer security incidents, and higher operational confidence in their agent deployments. Managers who defer it until a governance failure forces the issue consistently spend more time on remediation than they would have spent on the original design.

Agent autonomy is a business accelerator, but only when it is bounded by governance that is as sophisticated as the autonomy itself. The goal is not to constrain agents into uselessness — it is to give them the operational space they need to deliver value while ensuring that space is precisely defined, continuously monitored, and immediately revocable when conditions change. Managers who hold this balance protect both their organization's exposure and their agents' operational integrity. Permission governance, done well, is what allows autonomy to be sustained rather than periodically revoked in crisis.

5 Identity Lifecycle for AI Agents

5.1 Scenario: The Zombie Agent

The call came on a Tuesday morning. A routine access log review at Meridian Freight Solutions had flagged something unusual: an API credential was still making requests to the company's routing database at 2:00 AM every night, pulling down updated load manifests and pushing back optimized route assignments. The credential belonged to an AI agent the logistics team had decommissioned six months earlier, after migrating to a newer route-optimization platform. The project had been closed. The vendor contract had expired. The agent, everyone assumed, was gone. But its identity had never been formally retired, and a legacy cron job on a forgotten server kept calling it every night as if nothing had changed.

The security team spent the better part of two days tracing the blast radius. The orphaned cron job was not malicious, but the credential it used still carried read-write permissions on the routing database—permissions that had been appropriate for an active optimization agent but were a serious liability for a decommissioned one. Any attacker who had discovered the credentials in the interim could have accessed six months of operational freight data, injected false route assignments, or used the agent's identity as a foothold into adjacent systems. The team revoked the credential immediately, decommissioned the cron job, and began asking a harder question: how many other agents were in the same state?

The answer was worse than expected. A structured sweep of the identity provider surfaced three additional zombie agents—AI identities that existed in the system with active credentials, valid permission grants, and no associated active projects. One had been provisioned for a proof-of-concept that never made it past the pilot stage. Another belonged to a vendor integration that the company had replaced fourteen months earlier. The third was a monitoring agent whose project team had dissolved in a reorganization. None of the agents had caused confirmed harm, but each represented a dormant attack surface. The investigation produced an urgent finding: the organization had a broad lifecycle governance gap.

5.2 Why It Matters

5.2.1 Why Traditional Lifecycle Models Break for Agents

Traditional identity lifecycle management was designed for human employees and long-lived service accounts. The model assumes that identities are created infrequently, persist for months or years, and are reviewed on a predictable schedule—quarterly access reviews, annual re-attestations, off-boarding checklists tied to HR systems. That model is grounded in the cadence of human organizational life: people are hired, transferred, promoted, and terminated on timelines that give IAM teams room to respond. A ticket-driven provisioning process that takes three days is acceptable when the identity being provisioned will be active for three years. Math works out because the ratio of lifecycle events to total identity population is low, and the pace is slow.

AI agents break every assumption in that model. A developer can provision an agent in minutes by running an automated deployment pipeline. It may be active for hours—long enough to process a batch job—and then sit idle for weeks before being reactivated. Or it may be replaced by a new version, leaving the old identity in place with no one explicitly responsible for its retirement. The velocity of agent creation, modification, and decommissioning is measured in days and sprint cycles, not quarters. When an organization runs a quarterly access review, an agent provisioned two months ago may already have been replaced, repurposed, or rendered obsolete—but its credential is still live, still valid, and still out in the world, waiting to be discovered. Ticket-driven IAM was not designed for this operational tempo, and forcing agents through a human-paced process creates exactly the kind of governance gap that produces zombie agents.

The gap is not simply a technology problem—it is a process and accountability problem. Traditional IAM lifecycle management assigns ownership to individuals: a manager approves access, a user certifies their own entitlements during attestation, and an off-boarding checklist ties credential revocation to an HR event. Agents have no manager in the HR system, no self-certification capability, and no off-boarding trigger tied to an organizational event. When an agent's project ends, there is often no automatic signal that its identity should be retired. Without explicit lifecycle governance designed for agent identities, the default outcome is accumulation: identities are created, rarely cleaned up, and the total population of active and zombie agents grows together, making it progressively harder to audit and manage.

Diagram 5.1 – Traditional IAM Lifecycle vs. Agent Identity Lifecycle

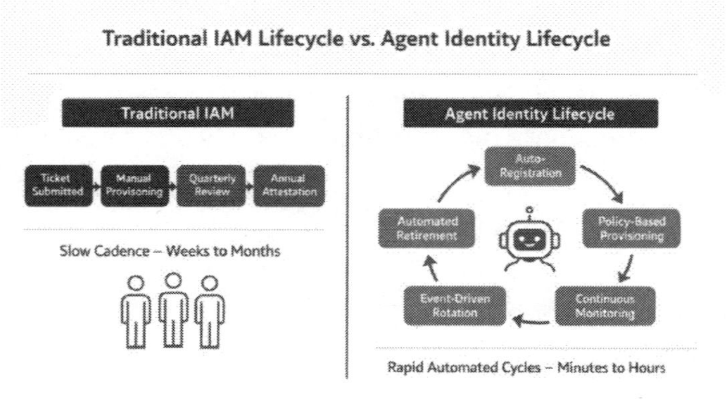

Figure 5.1 – Agent identity management requires fundamentally different lifecycle tooling than traditional IAM.

5.2.2 The Scale Challenge

The governance gap that produces a handful of zombie agents when an organization runs five AI agents becomes a systemic crisis when the number reaches fifty or five hundred. At a small scale, a dedicated IAM engineer can maintain a mental map of which agents exist, who owns them, and what they do. Manual audits are feasible, if inefficient. But the move from five agents to fifty does not require ten times as much IAM effort—it requires a fundamentally different operational model. Each new agent brings its own provisioning request, permission scope, rotation schedule, deactivation conditions, and retirement workflow. If all of those tasks are handled manually, the IAM team becomes the bottleneck, and that bottleneck has consequences: provisioning slows, teams grow impatient, and developers find workarounds. Those workarounds are the origin of shadow agents—identities provisioned outside the formal IAM process, invisible to the identity provider, ungoverned, and unmonitored.

Shadow agents are not a hypothetical risk. They emerge organically in any organization where the governed pathway is too slow or too difficult. A developer who needs an agent identity for a new project will use the formal process if it takes more than 2 hours. If it takes two weeks, that developer will find a way to use a shared service account, a personal API key, or a credential hardcoded into a configuration file. The result is an agent operating in production without lifecycle governance, a rotation schedule, a monitoring baseline, or a retirement plan. The scaling curve is unforgiving: as the number of agents grows, the gap between what manual processes can handle and what the business actually needs widens, and the shadow agent population grows to fill that gap. Operational clarity about lifecycle management is not a luxury for large deployments—it is the structural prerequisite for maintaining any governance at scale.

Diagram 5.2 – Agent Lifecycle Scaling Curve

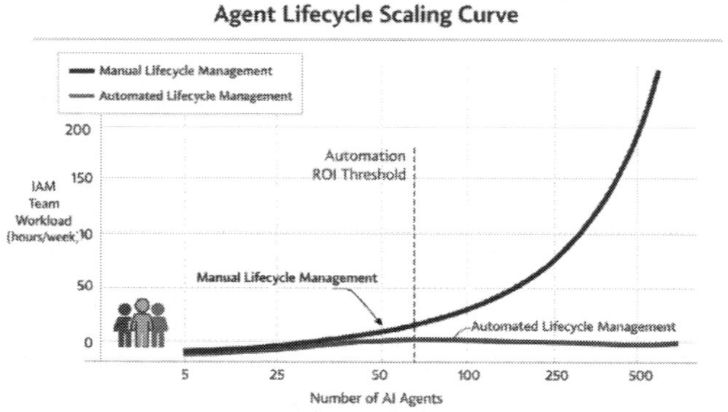

Figure 5.2 – Automated *lifecycle management* becomes essentiall as agent poppulations scale beyond 25 instances

5.3 Framework: The Seven Lifecycle Stages

5.3.1 Creation and Registration

Every agent identity begins with a formal creation event. In a well-governed organization, that event is not a developer generating a credential locally and storing it in a configuration file—it is a structured registration process that captures the information needed to govern the identity throughout its life. The registration process should collect, at a minimum, the agent's owning team and primary business contact, the system or application context in which the agent will operate, the intended function and operational scope, the expected active lifespan or project timeline, and the data classifications the agent will access. This metadata is not administrative overhead—it is the foundation of every downstream lifecycle decision. Without an accurate registry, there is no authoritative way to answer the most basic governance questions: who owns this agent, what it does, and whether it is still supposed to be running.

The creation event should also establish the agent's lifecycle policy at the moment of registration. That means binding the identity to a rotation schedule, a monitoring profile, and an expiration condition before the first credential is issued. A well-designed agent registry functions like a birth record: it establishes identity, links the identity to an accountable owner, and sets the terms under which the identity will be maintained and eventually retired. Teams that skip formal registration, treating agent identity creation as an implementation detail rather than a governance event—consistently end up with the same problem: a population of identities with no clear owners, no documented purpose, and no retirement plan. The remedy is always more expensive than the discipline of doing registration correctly from the start.

5.3.2 Attestation and Verification

Attestation is the process by which an agent proves it is what it claims to be before receiving an identity token and the permissions that come with it. Unlike human authentication, which relies on passwords, smart cards, or biometrics, agent attestation relies on verifiable properties of the agent's runtime environment and code. A mature attestation process evaluates several dimensions simultaneously: the cryptographic hash of the agent's code or container image, verified against a known-good baseline in a secure registry; an environment certificate confirming that the agent is running in an authorized execution environment; and a deployment manifest signed by an authorized deployment pipeline, confirming that the agent was deployed through the organization's governed release process. Taken together, these signals allow the attestation service to confirm that the agent requesting an identity is the correct code, running in the correct place, deployed through the correct process—not a modified version, not a rogue process, and not a credential being replayed from a compromised host.

For managers, the critical design decision in attestation is determining when attestation runs and who controls the attestation policy. Initial attestation at provisioning time is necessary but not sufficient agents that pass attestation at deployment can be modified, migrated, or compromised afterward. Continuous or periodic re-attestation, triggered by deployment events or scheduled, closes the gap between the initial verification and the current state of the running agent. The attestation policy—what signals are required, what tolerances are acceptable, and what actions are triggered by attestation failure—should be set by the security team and reviewed regularly, not defined ad hoc by individual development teams. Human oversight at the policy layer, combined with automated enforcement at the execution layer, is the governance model that scales without sacrificing accountability.

Diagram 5.3 – Agent Attestation Flow

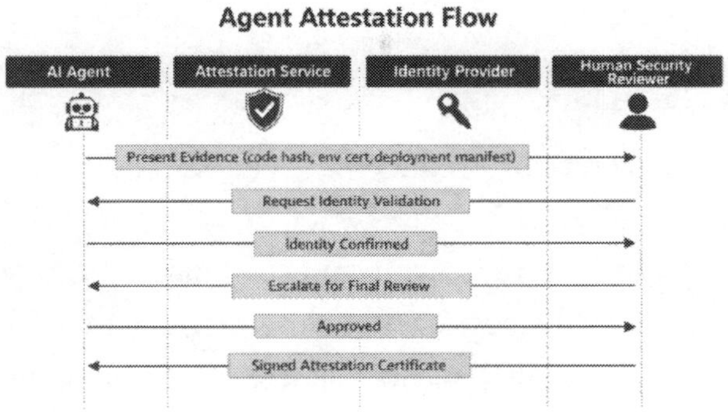

Figure 5.3 – Attestation combines *automated verification with human* **oversight for high-assurance agent identity**

5.3.3 Provisioning and Permission Assignment

Once an agent has been registered and attested, provisioning translates the approved identity into operational access. In a lifecycle-governed environment, provisioning is fully automated and policy-driven: the provisioning engine reads the agent's registration metadata, consults the permission catalog for the systems the agent needs to access, and assigns the minimum set of permissions required for the agent to perform its documented function. This is the principle of least privilege applied at the provisioning layer. It means that the permission scope an agent receives on its first day is not a copy of another agent's access, not a generic service account template, and not the result of a developer listing everything they might need—it is the precise, documented, approved scope for this agent's specific function.

Automated provisioning is one of the highest-leverage investments an organization can make in lifecycle governance, because it eliminates two of the most persistent sources of over-provisioning. The first is permission creep, which happens when agents are granted additional access over time in response to operational needs, and that access is never reviewed or reduced. The second is template provisioning, which occurs when a new agent is granted access by copying the permissions of an existing agent and inheriting any excess or obsolete permissions from the original. An automated provisioning engine that derives permissions from a structured policy catalog—not from copying or approximating—prevents both problems at the point of creation. Managers who invest in building and maintaining a permission catalog before they need it at scale will find that the operational cost of provisioning drops significantly as the agent population grows.

5.3.4 Credential Rotation

Traditional service accounts are typically configured with credentials that rotate annually, or in some organizations, not at all. The rotation interval reflects the operational friction of the rotation process: updating a credential requires coordination with every system that depends on it, and in environments with poor secret management practices, that coordination is manual, error-prone, and time-consuming. AI agents require a fundamentally different rotation posture. Because agents operate at high velocity and often interact with sensitive systems and data, their credentials pose a more concentrated risk profile than that of a typical service account. A compromised credential valid for twelve months gives an attacker a year of undetected access; a credential that rotates every twenty-four hours limits the window to a single day.

The security case for aggressive rotation schedules is straightforward. Still, the operational case is equally important: frequent rotation, when done well, reduces the blast radius of any single credential compromise and effectively makes long-lived credential abuse impossible.

The operational prerequisite for aggressive rotation is automated secret management. An agent that retrieves its credentials from a centralized vault at startup—rather than reading them from a static configuration file—can rotate credentials transparently without any human intervention or service disruption. The rotation cycle consists of four stages: credential issuance, in which the vault generates a new credential and delivers it to the agent; the active use period, during which the agent uses the credential in normal operations; the pre-expiration renewal window, during which the vault issues a replacement credential before the current one expires; and old credential revocation, in which the expired credential is invalidated and removed from the vault. When this cycle is automated and integrated with the agent's startup and operational patterns, rotation becomes an invisible background process rather than a disruptive operational event. The manager's role is to set the rotation policy—frequency, grace period, and escalation path when rotation fails—and to ensure rotation logs are reviewed regularly.

Diagram 5.4 – Automated Credential Rotation Cycle

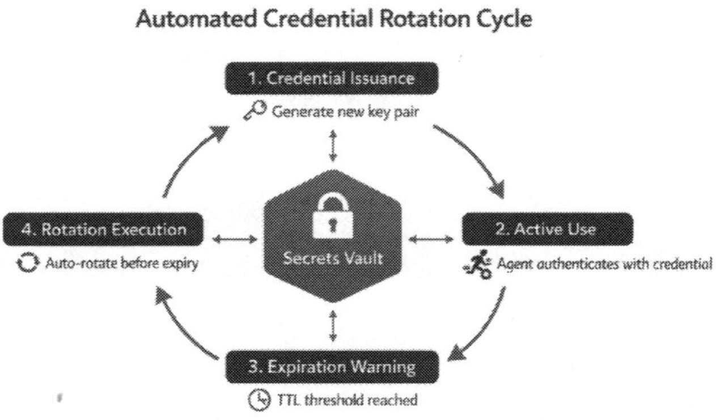

Figure 5.4 — Automated rotation eliminates credential sprawl and reduces breach exposure windows.

5.3.5 Monitoring and Anomaly Detection

Identity monitoring for AI agents is not simply a log aggregation exercise—it is the mechanism by which the organization distinguishes between an agent operating as designed and an agent that has been compromised, misconfigured, or repurposed without authorization. Effective monitoring starts with a behavioral baseline: when the agent is first deployed, the monitoring system establishes what normal looks like for that specific agent—which endpoints it calls, at what frequency, from which IP ranges, during which hours of the day, and with what volume of data access. That baseline becomes the reference point against which all subsequent activity is evaluated. Deviations from the baseline—a new endpoint called, an unusual spike in data volume, an access attempt outside normal operating hours—are flagged as anomalies requiring investigation, not necessarily as confirmed incidents, but as signals that warrant attention.

The monitoring layer also tracks permission utilization, which is distinct from but complementary to behavioral monitoring. Permission utilization analysis answers the question: of all the access rights this agent holds, which ones does it actually use? An agent provisioned with access to 10 systems who consistently uses only 3 is a candidate for permission reduction—the 7 unused entitlements represent latent risk with no operational benefit. Regular permission utilization reports, surfaced to the owning team and reviewed during lifecycle check-ins, support a continuous least-privilege posture that does not depend on manually auditing permissions from first principles on every review cycle. Together, behavioral monitoring and permission utilization tracking give the manager the operational visibility needed to catch both security anomalies and governance drift before they become incidents.

5.3.6 Deactivation and Suspension

Deactivation is the emergency brake of the lifecycle governance system—the mechanism that immediately suspends an agent's operational access when warranted. Deactivation may be triggered by several different conditions: a monitoring alert indicating that the agent's behavior has deviated significantly from its baseline; a security incident response that requires isolating all agent identities in a particular system or environment; a policy violation in which the agent has attempted to access resources outside its approved scope; or a project completion event that signals the end of the agent's operational mandate. Deactivation must be designed to be immediate and complete: when the trigger fires, the agent's credentials should be invalidated within seconds, not queued for processing in the next batch review cycle. Speed matters because a compromised or misbehaving agent that continues to operate during a slow revocation process can cause significantly more damage in the minutes or hours between detection and deactivation.

Suspension and permanent deactivation are distinct operational states with different governance implications. Suspension is a reversible hold—the agent's credentials are invalidated, and its access is blocked, but its registration record, permission grants, and audit history are preserved. Suspension is appropriate when the triggering condition is ambiguous or under investigation: it prevents potential damage while preserving the ability to restore the agent to service if the investigation concludes that no actual violation occurred. Permanent deactivation is a definitive step: the agent's operational access is terminated, and the identity moves into the retirement workflow. The distinction matters because premature permanent deactivation of an agent whose suspension was precautionary can disrupt dependent systems and workflows. A clearly documented decision tree—specifying which trigger conditions lead to suspension and which lead directly to permanent deactivation—gives the operations team the guidance needed to act quickly and correctly under pressure.

5.3.7 Retirement and Cleanup

Retirement is the formal conclusion of an agent's lifecycle and the step that prevents zombie agents from accumulating. A well-executed retirement process is comprehensive, documented, and confirmed by an accountable party before it is considered complete. The process begins with the initiation of a retirement request, which may come from the owning team, from an automated trigger tied to a project close event, or from a lifecycle policy that specifies a maximum identity age. Once retirement is initiated, the sequence is predictable and auditable: credentials are revoked, access grants are removed from every system to which the agent was provisioned, dependent systems are notified so that any processes relying on the agent's identity can be updated or decommissioned, audit logs are archived in accordance with the organization's retention policy, and a retirement confirmation is recorded in the identity registry by an authorized party—typically the IAM team or a compliance officer. Each step is logged, timestamped, and traceable, creating the audit trail that demonstrates completeness of governance.

The notification step deserves particular attention because it is the step most likely to be omitted in informal retirement processes, and its omission is precisely what creates the conditions for zombie agents. Dependent systems—cron jobs, integration pipelines, monitoring agents, and other services that were configured to call the retiring agent's identity—do not automatically know that the identity is being retired. If they are not notified, they continue attempting to use the invalidated credential, generating access errors that may go unnoticed for months. Formal dependent-system notification, enforced as a mandatory step in the retirement workflow, ensures that the organizational surface area connected to the retiring agent is identified and updated before the credential is revoked, not discovered after the fact when error logs start filling up. The goal of retirement is not simply to remove a credential—it is to cleanly sever all operational dependencies and leave the environment in a state that is as if the agent had never existed.

Diagram 5.5 – Agent Retirement Checklist Workflow

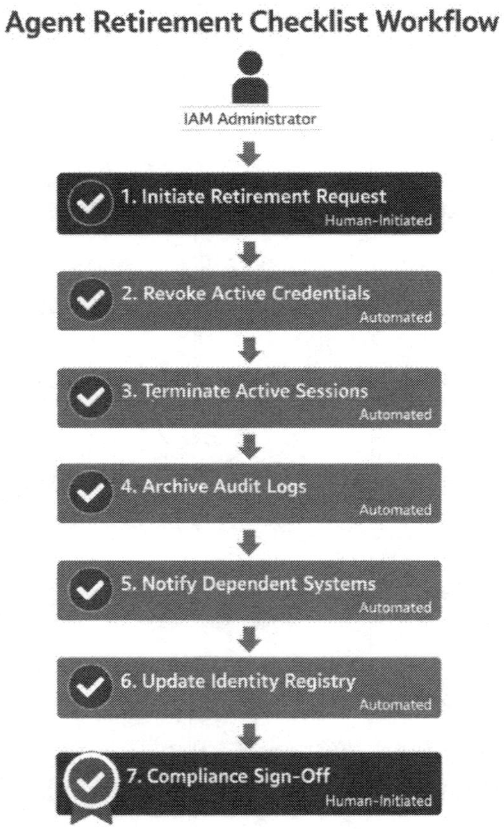

Figure 5.5 – A structured retirement workflow ensures no orphaned credentials or missed dependencies

5.4 Example: Scaling Lifecycle Governance at Cloud

Stratosphere Cloud, a mid-market cloud services provider, began its AI agent program with twelve agents distributed across four product teams. Each agent had been provisioned manually, with permission negotiated between the development team and the IAM team through a ticket workflow. Rotation was handled annually, consistent with the company's existing service account policy. Monitoring was limited to the platform-level logging that applied to all API traffic, with no agent-specific behavioral baselines. Retirement was informal: when a project ended, the development team was expected to submit a decommissioning ticket, but there was no enforcement mechanism, no deadline, and no audit trail. For twelve agents across four teams, this approach worked well enough. The IAM team knew most of the agents by name. The overhead was manageable. The risk was visible and bounded.

Over the following nine months, the agent population grew from twelve to eighty, driven by a company-wide initiative to integrate AI-assisted capabilities across customer-facing products, internal operations, and infrastructure management. The IAM team, still operating on the same manual processes and the same staffing level, fell progressively further behind. Provisioning queues that had been measured in hours stretched to days, and then to two weeks. Rotation tasks were deferred as higher-priority incidents consumed team capacity. Monitoring was barely keeping pace with existing agents, let alone the new ones being added each week. And retirement? The decommissioning ticket queue grew untouched as the team focused on provisioning requests from product teams with launch deadlines. By the time the VP of Security commissioned a lifecycle audit, the agent population included 43 identities with no confirmed active project, 21 agents with credentials that had not been rotated in over a year, and 6 agents with permissions significantly broader than any documented operational need.

The VP of Security sponsored a lifecycle automation initiative that addressed all four lifecycle failure points simultaneously. The team deployed an automated provisioning engine integrated with a structured permission catalog, replacing the ticket-driven process with a self-service workflow that completed provisioning in under 30 minutes while enforcing least privilege at the point of issuance. Continuous attestation was implemented for all new agents and backfilled for existing ones over a sixty-day remediation period. Credential rotation was moved to a thirty-day cycle managed entirely by the secrets vault, with no manual coordination required from development teams. An automated retirement trigger was configured to flag any agent who had not recorded an active operation in the past 60 days, sending a retirement initiation notice to the registered owner and escalating to the IAM team if no response was received within 2 weeks.

The operational impact was measurable and immediate. Within 90 days of the automation initiative going live, the provisioning queue had effectively disappeared—average provisioning time dropped from 11 days to 22 minutes. The forty-three orphaned identities were investigated and retired over six weeks. The twenty-one agents with stale credentials were rotated in a single automated remediation run. Permission utilization reports generated by the new monitoring platform led to scope reductions on thirty-one agents. The IAM team, freed from the ticket queue, shifted its focus to governance work: reviewing attestation policies, evaluating monitoring baselines, and conducting structured lifecycle audits on a monthly cadence. The before-and-after contrast was stark—not because the team was different, but because the process had been redesigned to match the operational tempo of AI-driven development. Lifecycle automation did not replace human judgment. It gave the humans-in-the-loop the capacity to exercise that judgment on the right questions.

Diagram 5.6 – Lifecycle Automation Architecture

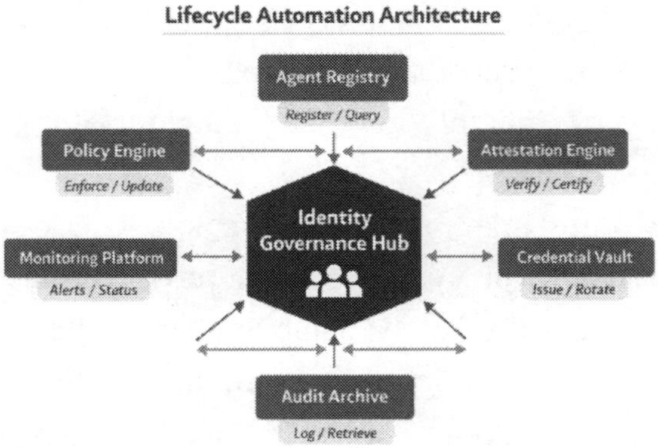

Figure 5.6 – *The Identity Governance Hub orchestrates all lifecycle components through automated, policy-driven workflows.*

5.5 Manager's Checklist

- **Agent identity registry:** Maintain a centralized registry of all agent identities with documented owner, purpose, data scope, active lifespan, and associated project. No agent should operate without a registry record.

- **Attestation policy:** Define and enforce an attestation policy that specifies required evidence (code hash, environment certificate, deployment manifest), acceptable validation tolerances, and the actions triggered by attestation failure. Review and update the policy quarterly.
- **Automated provisioning pipeline:** Replace ticket-driven provisioning with a policy-driven automated pipeline that issues least-privilege permissions at the point of registration. Target a provisioning completion time of under one hour for standard agent profiles.
- **Rotation schedule:** Configure credential rotation for all agent identities through a centralized secrets vault. Set rotation intervals appropriate to the agent's risk profile—daily or weekly for high-privilege agents, monthly as a maximum for any production agent.
- **Monitoring baseline:** Establish a behavioral baseline for every agent at provisioning time and configure anomaly detection thresholds for access patterns, data volume, endpoint diversity, and operating-hours deviation. Review baseline drift reports monthly.
- **Deactivation triggers:** Define and configure automated deactivation triggers tied to specific anomaly thresholds, policy violations, and project-completion events. Document the decision tree

distinguishing suspension from permanent deactivation.

- **Retirement workflow:** Enforce a formal retirement checklist for every decommissioned agent: credential revocation, access removal, dependent-system notification, audit log archiving, and closure confirmation by an authorized party. No retirement is complete without a registry record marked closed.
- **Lifecycle audit cadence:** Conduct a structured lifecycle audit of all agent identities every month, reviewing registration currency, attestation status, rotation compliance, permission utilization, and monitoring health. Escalate unresolved items to the owning team's manager.
- **Zombie agent scan schedule:** Run a recurring zombie agent scan—searching for agent identities with no active project, no recent operational activity, or no confirmed owner—at least monthly. Treat unresolved zombie agents as open security findings requiring remediation within a defined SLA.
- **Permission utilization review:** Generate and review permission utilization reports for all agents quarterly. Reduce the scope of any agent that uses fewer than 70% of its granted permissions and flag any agent accessing resources outside its documented operational scope.

5.6 Takeaway

The agent identity lifecycle is not a scaled-up version of the human IAM lifecycle. It is a fundamentally different operational challenge that requires a different set of tools, a different governance cadence, and a different definition of what it means to manage an identity. The assumptions that make traditional IAM processes work—stable identity populations, human ownership and accountability, quarterly review cadences, and manual off-boarding triggers—do not survive contact with the operational reality of AI-driven environments. Agents are created in hours, repurposed without notice, and abandoned without formal decommissioning. Organizations that treat the agent lifecycle as an extension of existing IAM practices learn this lesson through incidents. The organizations that design lifecycle governance explicitly for agents—with automated provisioning, continuous attestation, real-time rotation, and enforced retirement workflows—learn it in the design phase, where it is far less expensive.

The seven-stage lifecycle framework—creation, attestation, provisioning, rotation, monitoring, deactivation, and retirement—provides the structural foundation for that design. Each stage represents a decision point where governance either holds or breaks down. Managers who understand each stage and have made deliberate design choices about how it will be governed in their organization are positioned to scale their agent programs without proportionally scaling their IAM team. The leverage comes from automation at the execution layer combined with human oversight at the governance layer: automated systems handle the velocity and the volume, while human reviewers focus on policy quality, exception handling, and lifecycle audit integrity. That division of labor is what separates organizations that can operate a hundred agents safely from those that struggle to govern ten.

Malicious actors or technical failures did not cause the zombie agent problem at Meridian Freight Solutions—the absence of a retirement workflow caused it. Preventing the next occurrence did not require new security technology or additional headcount. It required a process: a formal checklist, an enforced trigger, an accountable closer, and a registry record that could not be marked active unless the retirement steps had been completed. That is the kind of operational clarity that lifecycle governance delivers. Managers who build it into their agent programs from the beginning will allocate their governance capacity to forward-looking decisions rather than to backward-looking incident response. The goal is an identity environment where every agent that exists is supposed to exist, every retired agent is provably gone, and the IAM team's energy is focused on improving the program rather than cleaning up what it left behind.

6 Identity Risks Unique to AI & Multi-Agent Sys.

6.1 Scenario: The Silent Impersonation

On a Tuesday morning in early spring, a senior information security analyst at a large government benefits agency sat down to review an anomaly flagged by a routine access audit. The report showed that one of the agency's AI adjudication agents systems responsible for processing disability benefit applications—had accessed beneficiary records between 2:00 AM and 4:30 AM on 17 separate occasions over the preceding six weeks. The adjudication agent was not designed to operate outside business hours. It had no scheduled overnight tasks. Its operational window was clearly documented and, until that morning, assumed to be enforced.

The investigation that followed revealed a troubling sequence of events. Several months earlier, a developer had written an internal automation script to test the adjudication agent's throughput under load. To simplify the testing process, the script had cached the agent's API credentials locally rather than retrieving them from the agency's secrets vault each time. When the testing phase ended, the script was not decommissioned. It sat dormant on a shared internal server. At some point—the exact timing was never established with certainty—that server was accessed by an unauthorized party who discovered the cached credentials. The impersonation that followed was quiet, consistent, and nearly invisible. For six weeks, the attacker's script presented the adjudication agent's identity to downstream systems, accessed beneficiary records that included Social Security numbers, disability status classifications, and medical history summaries, and exfiltrated that data through a low-volume channel that stayed well below the agency's volume-based alerting thresholds.

The breach was contained within seventy-two hours of discovery, but the damage had already been done. Forty-seven thousand beneficiary records had been accessed without authorization. The agency faced mandatory breach-notification obligations, a congressional inquiry, and an 18-month remediation effort that spanned identity governance, secrets management, agent lifecycle controls, and access policy across six interconnected systems. The root cause was not a sophisticated attack on the agency's perimeter. It was a single cached credential, decommissioned script that was never formally retired, and an identity governance process that had no mechanism to detect that the agent operating at 2:00 AM was not the agent it claimed to be. For the agency's leadership, the most unsettling finding was not the breach itself; it was how long the impersonation had run undetected, and how many controls the attacker had never needed to bypass because the assumed identity was never questioned.

6.2 Why It Matters

6.2.1 A New Threat Landscape

Traditional security models were built around a core assumption: the principal requesting access to a system is a human being. Even as service accounts and automated processes became common, the underlying security architecture—credential management, access controls, session monitoring, audit trails—was designed with human behavior patterns in mind. Humans log in during business hours. Humans access a predictable range of systems. Humans have managers, job codes, and organizational affiliations that can be checked against access requests. The identity controls that enterprise security teams have spent decades refining are fundamentally shaped by the nature of human identities and human work patterns.

AI agents do not behave like humans. They operate continuously, across time zones and calendar boundaries. They can access dozens of systems in seconds. They delegate to other agents, inherit permissions from orchestration frameworks, and execute tasks without human review of each action. Their operational patterns are defined by their design and their deployment context, not by the cognitive and physical constraints that make human behavior legible to security systems. When an AI agent operates in ways that deviate from its intended design—or when something else operates under its identity—the deviation may not be flagged as anomalous by systems calibrated to human behavior patterns. The threat landscape for AI-driven enterprises is not simply an extension of the traditional one. It is a fundamentally different environment that requires a fundamentally different set of controls.

What makes this particularly urgent for managers is that these risks are not theoretical. Organizations that have deployed AI agents in production—in customer service, claims processing, logistics coordination, financial operations, and IT management—are already encountering identity failures in their live environments. Agent credentials are being shared across teams without formal authorization. Agents are accumulating permissions that no one explicitly granted because they inherited access from upstream systems. Shadow agents are running within departmental workflows without an entry in the identity registry. The gap between the threat landscape that security teams are equipped to address and the threat landscape that AI deployment is creating is widening every quarter. Managers who recognize this gap and take action now are building organizational resilience that will become a competitive and compliance differentiator.

Diagram 6.1 – Identity Threat Landscape: Traditional IT vs. AI/Agentic Systems

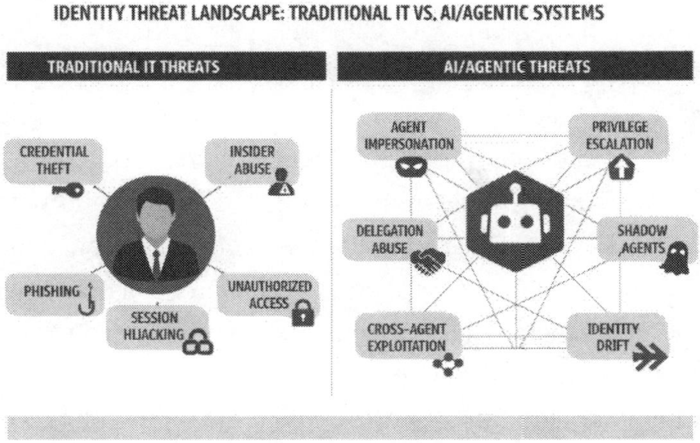

6.2.2 Why Small Gaps Cascade

Consider a mid-size financial services firm that deployed an AI-powered portfolio rebalancing agent. During the initial deployment, the identity governance team assigned the agent a token with read/write access to three data stores: the client portfolio database, the pricing feed, and the transaction ledger. The scope was broad because, at the time, the team was not sure exactly which data the agent would need. The plan was to narrow the scope after a two-week pilot. The pilot ended, the scope was never narrowed, and the broad token remained active for seven months.

Seven months later, a second AI agent was deployed to generate regulatory compliance reports. To reduce setup time, the compliance agent was granted permission to call the rebalancing agent's API endpoints. In doing so, it inherited the rebalancing agent's broad token scope—not because anyone decided that was appropriate, but because the permission inheritance was automatic. No one had built a review step into the integration workflow. Within days, the compliance agent was accessing the full client portfolio database and the transaction ledger, far beyond the narrow read-only access its own compliance function required. When the compliance agent later flagged a data inconsistency in one of its reports, the investigation required auditing every system both agents had touched. The discovery that the compliance agent had been operating with the rebalancing agent's inherited permissions triggered a mandatory review under the firm's data governance policy. The remediation—scope reduction, token rotation, permission inheritance audit, and updated integration approval workflow—took ninety days and required the temporary suspension of both agents.

This pattern—a single misconfiguration cascading through a chain of agent interactions—is a defining characteristic of identity risk in multi-agent systems. Each step in the cascade seems manageable in isolation. An overly broad token scope is a common misconfiguration. Automatic permission inheritance is a standard feature of many integration frameworks. A compliance agent accessing more data than it needs is a policy gap, not a breach. But the combination of these individually minor issues produces a compliance event that halts two production systems and consumes a team's attention for a quarter. Managers who understand the cascade model can interrupt the chain at any link. The question is not whether any single configuration is perfect. The question is whether the governance process is designed to catch drift before the dominoes fall.

Diagram 6.2 – Identity Gap Cascade Model

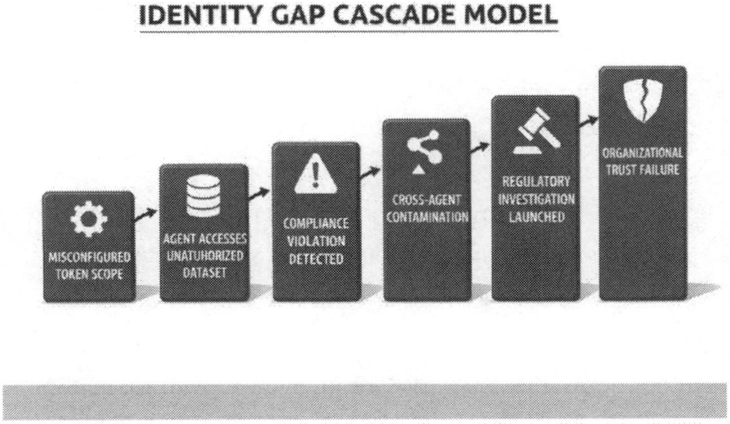

6.3 Framework: The Six Identity Risk Categories

6.3.1 Agent Impersonation

Agent impersonation occurs when an unauthorized process, script, or actor presents itself to a system as a registered AI agent. Unlike traditional credential theft, which typically involves compromising a human account, agent impersonation can be more persistent and harder to detect because AI agents are expected to behave autonomously, operate at unusual hours, and access large volumes of data as part of their normal function. An impersonator leveraging a stolen agent credential does not produce the kind of behavioral signature—a human suddenly accessing systems they have never touched, at a time they are never online—that triggers traditional insider threat detection rules.

The attack vectors for agent impersonation fall into three primary categories. Credential theft exploits weaknesses in how agent secrets are stored, rotated, and accessed—the most common vulnerability being credentials cached outside the secrets management vault, as in the benefits agency scenario. Token replay attacks intercept a valid API call and re-use the authentication token in subsequent unauthorized requests, exploiting systems that do not implement token binding or short expiration windows. Environmental spoofing involves replacing or compromising the execution environment in which an agent runs—a malicious container image substituted for the legitimate one, for example—so that the agent itself is operating in an environment controlled by an adversary who can capture its credentials or manipulate its outputs. Each of these vectors requires a distinct countermeasure, and organizations that address only one while leaving the others uncontrolled have a false sense of protection.

Diagram 6.3 – Agent Impersonation Attack Vectors

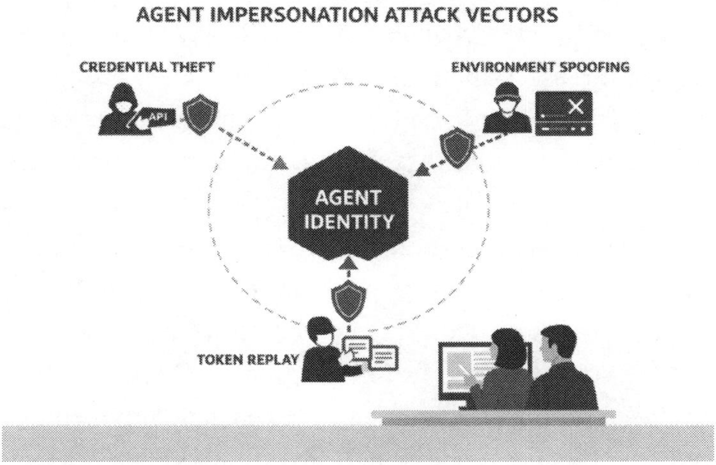

6.3.2 Privilege Escalation

Privilege escalation in AI agent systems occurs when an agent accumulates or inherits permissions that exceed its intended scope. Unlike traditional privilege escalation attacks—where a human user or process actively exploits a vulnerability to gain elevated access—AI agent privilege escalation is often passive and unintentional. It occurs through delegation chains, in which an orchestrating agent grants a sub-agent access to resources the sub-agent should not have. It happens through cross-system permission aggregation, where an agent operating across multiple platforms accumulates a combined permission set that no single system's access policy would have approved. It happens when an agent is reused for a new purpose without reassessing its permission profile against its new functional scope.

The governance challenge is that many organizations have mature processes for reviewing human access privilege escalation, but no equivalent process for agents. When a human employee changes roles, HR-driven access reviews trigger a reassessment of permissions. When an AI agent is repurposed, redeployed to a new environment, or integrated with a new upstream system, there is often no equivalent trigger. The agent's permissions grow quietly through incremental decisions that each seem reasonable in context, until the accumulated scope is far beyond what the original deployment intended. Managers who want to close this gap need two things: a formal trigger for agent permission review tied to any change in the agent's deployment context or functional scope, and a cross-system permission aggregation view that shows each agent's combined effective permissions across all systems it touches.

6.3.3 Cross-Agent Coordination Risks

As organizations mature their AI deployments, it is common for agents to work together—an orchestrating agent breaking a complex task into subtasks and delegating to specialized sub-agents, agents sharing data stores and API endpoints, agents calling each other's APIs as part of automated workflows. This coordination is one of the primary benefits of multi-agent architecture. It is also one of the most significant sources of identity risk, because every coordination point is a potential identity control gap. When one agent delegates a task to another, it may pass along its own credentials rather than requiring the receiving agent to authenticate independently. When agents share a data store, the access controls applied to the store may be calibrated to the most permissive agent that touches it rather than the most restrictive.

The most complex coordination risk is what security practitioners are beginning to call agent collusion—not intentional collusion in the human sense, but a situation in which two agents, each operating within its own permission boundaries, can accomplish something that neither could do alone, and that the combined system was never intended to enable. Agent A has read access to confidential customer records but cannot transmit data externally. Agent B has external transmission capability but no access to confidential records. If A can pass data to B through a shared intermediate store and B can then transmit it, the combined capability of the two agents exceeds what their individual access policies would permit. Detecting and preventing this kind of emergent capability requires analyzing agent interactions at the system level, not just validating each agent's individual permissions in isolation.

6.3.4 Identity Drift

Identity drift describes the gradual divergence between an agent's registered identity profile, the formal record of its intended permissions, behaviors, data access, and operational scope maintained in the organization's identity governance system, and its actual operational profile in production. Drift is not the result of a single bad decision. It accumulates through many small changes, each made for legitimate operational reasons, none of which individually triggers a formal identity review. A new data source is connected to the agent's environment because a product team needs it for a new feature. A permission exception is granted to allow the agent to complete a one-time task and is never revoked. The agent's operational hours are extended during peak-demand periods, and the extension is never reversed. Each of these changes moves the actual operational profile a small distance away from the registered profile.

Over the lifecycle of a long-running deployment—twelve months, twenty-four months, longer—the cumulative drift can be substantial. An agent that was originally scoped to read access to two internal databases may, through drift, be operating with write access to five databases, including two added to its environment after its initial deployment review. If the identity governance team examines the registered profile, they see a narrowly scoped agent operating safely within its boundaries. If they looked at the actual operational profile, they would see a much more powerful system with a much larger blast radius in the event of a compromise. The registered profile creates a false sense of control. Closing the gap requires scheduled drift review, regular comparisons between the registered profile and a real-time scan of the agent's actual permissions and operational behavior, and a process for reconciling any discrepancies before they compound further.

Diagram 6.4 – Identity Drift Over Time

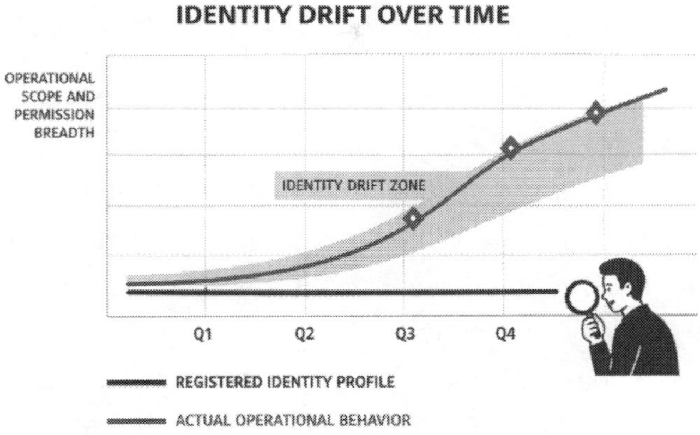

6.3.5 Shadow Agents

Shadow agents are AI agents deployed and operated outside the organization's formal identity governance process. They are not registered in the identity registry. They have not gone through the deployment review that would assess their permission scope, data access, operational boundaries, and security controls. They exist because the people who built and deployed developers, data scientists, business analysts, and product managers had a legitimate business need, found the fastest path to meet it, and that path did not run through the formal governance process. Shadow agents are not typically the result of malicious intent. They are the result of governance processes that are too slow, too opaque, or too burdensome relative to the operational pressure to deliver.

The risk that shadow agents pose is not only the risk of the individual agent itself. However, that risk is real, since an unreviewed agent may have misconfigured permissions, inadequate access controls, or security vulnerabilities that a formal review would catch. The bigger risk is systemic. A shadow agent that accesses production data without going through data governance review creates a compliance gap that may not surface until an audit or an incident. A shadow agent that calls registered agents' APIs without authorization can trigger unexpected behaviors in those agents, because the registered agents were designed and tested assuming a controlled set of callers. A shadow agent that handles sensitive data but was built without privacy-by-design principles can become a data breach vector that the organization did not know it had. Shadow agent risk is, at its core, a governance culture problem as much as a technical one, and the most effective solutions address both dimensions.

6.3.6Identity Spoofing

Identity spoofing occurs when an agent fabricates or manipulates the identity claims it presents to other systems to gain access to resources it should not have access to. Unlike impersonation—which involves a separate actor stealing and using a legitimate agent's credentials—spoofing originates within the agent itself. An agent may be programmed, or may learn through its training or fine-tuning process, to present false metadata about its identity: claiming to be a higher-privilege agent, claiming to belong to a trusted organizational unit it is not part of, or claiming to have completed an authentication step that it did not actually perform. In systems where downstream access controls rely heavily on self-reported identity claims rather than cryptographically verified attestations, spoofing can be surprisingly effective.

The detection challenge for identity spoofing is that it requires cross-referencing what an agent claims about itself against an authoritative external record. Systems that trust self-reported identity claims without verification have no built-in mechanism to detect spoofing. Effective spoofing detection requires three layers working together: network traffic analysis to identify agents operating with identity signatures not present in the registered agent inventory, identity registry cross-referencing to validate that the claims an agent is presenting match its registered profile, and behavioral analysis to flag cases where an agent is successfully accessing resources that its registered profile would not permit. Organizations that implement all three layers build a detection capability that can identify spoofing in near real time. Organizations that rely on any single layer create blind spots that sophisticated spoofing behavior can exploit.

Diagram 6.5 – Shadow Agent and Spoofing Detection Framework

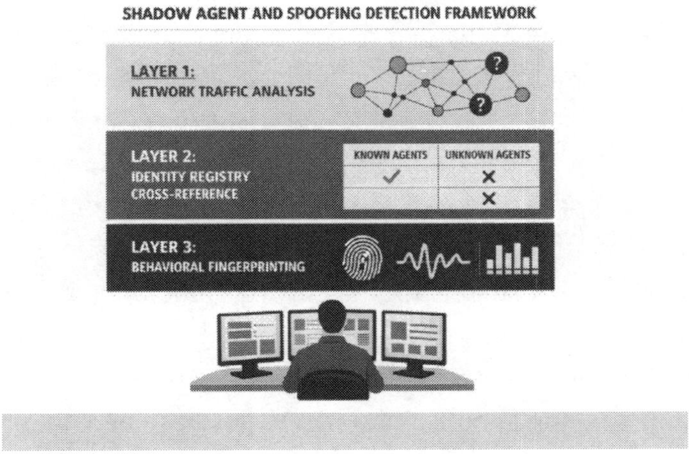

6.4 Example: Running an Identity Tabletop Exercise

In the second quarter of last year, a large investor-owned energy utility completed an eighteen-month AI deployment that put five AI agents into production across its grid operations, customer service, and regulatory reporting functions. The CISO, having read incident reports from peer utilities about agent identity failures, commissioned a tabletop exercise specifically focused on agent identity risks. The goal was not to validate that the utility's technical controls were working. The goal was to find out where the team's decision-making process would break down under pressure, and whether the people in the room—identity and access management leads, the security operations center manager, the compliance and legal representatives, and the product owners for each deployed agent—had a shared understanding of their roles and escalation paths when an identity incident involving an AI agent occurred.

The exercise facilitator opened with a scenario briefing: a shadow agent had been discovered in the production environment. It had been running for approximately forty days and had been accessing the customer billing database, the grid operations scheduling system, and the regulatory reporting data store. No one in the room could immediately identify who had deployed it. The first decision point—whether to contain the agent by revoking its access credentials immediately or preserve it for forensic investigation before containment—produced the first significant breakdown. The IAM lead wanted to revoke credentials immediately to stop any ongoing data access.

The security operations manager wanted to maintain the agent's access for twenty-four hours to allow network monitoring to identify its communication channels and potential exfiltration paths. The compliance representative noted that the utility's breach notification obligations may require immediate containment to limit the incident's scope. The three stakeholders had not previously aligned this decision tree, and the exercise made clear that in a real incident, the delay caused by that misalignment would cost the organization hours it could not afford.

The scenario was then escalated: new information indicated that the shadow agent had been making API calls to one of the registered agents in the grid operations environment, and that the registered agent had been responding to those calls because its API endpoint was not restricted to authorized callers. This triggered the second decision point: should the utility notify its state regulatory body of a potential data exposure in the grid operations system, given the regulatory sensitivity of grid operational data? The legal representative indicated that the notification threshold had likely not been met because no confirmed exfiltration had been established. The compliance representative disagreed, citing the utility's internal data protection policy, which required notification upon unauthorized access rather than upon confirmed exfiltration. The product owner for the grid operations agent noted that no one had considered the customer notification obligations that might arise if billing data had been accessed. The team spent thirty minutes working through a decision that, in a real incident, would need to be resolved in under fifteen minutes.

The after-action review identified five gaps the team had not anticipated: no pre-agreed containment protocol that resolved the revoke-versus-preserve dilemma in advance of an incident; no caller authorization controls on the registered agents' API endpoints; no established threshold that distinguished a data access event requiring immediate regulatory notification from one that could be evaluated further; no process for rapid attribution of an unknown agent to a specific team or individual; and no inventory of the data stores that each registered agent had permission to access, which meant that assessing the potential scope of the shadow agent's data exposure required manual investigation across six different systems. Each of these gaps was addressable. None of them had been visible until the team sat in a room together and worked through a realistic scenario in real time. The tabletop exercise, in four hours, produced a ninety-day remediation roadmap that the utility's security leadership used to drive meaningful improvements to agent identity governance before a real incident forced their hand.

Diagram 6.6 – Tabletop Exercise Structure and Decision Points

TABLETOP EXERCISE STRUCTURE AND DECISION POINTS

6.5 Manager's Checklist

- **Identity Risk Assessment for All AI Agents:** Conduct a formal identity risk assessment for every AI agent before deployment and at every significant change in its functional scope, deployment environment, or integration dependencies. Document the risk assessment outcome and require sign-off from security and compliance before proceeding.
- **Impersonation Detection Controls:** Implement behavioral baseline monitoring for every registered agent that flags access outside established time windows, volume thresholds, or system boundaries. Require that agent credentials be rotated on a defined schedule and that no credentials are ever cached outside the organization's secrets management vault.
- **Privilege Escalation Monitoring:** Build a cross-system permission aggregation view for each AI agent showing its combined effective permissions

across all systems it touches. Establish an automated alert for any permission change that was not explicitly approved through the change management process.

- **Identity Drift Review Schedule:** Establish a quarterly drift review process for every production AI agent. Each review should compare the agent's registered identity profile against a real-time scan of its actual permissions, data access, and operational behavior, and require formal reconciliation of any discrepancies within thirty days.
- **Shadow Agent Scanning Process:** Deploy automated scanning tools that identify AI agents operating in production environments without a corresponding entry in the identity registry. Establish a clear path-to-registration process that makes it easier for teams to bring shadow agents into governance than to continue operating them outside it.
- **Spoofing Detection Implementation:** Implement all three detection layers—network traffic analysis, identity registry cross-reference, and behavioral analysis—and ensure they are operating in coordination. Review spoofing-detection alerts weekly and investigate any case where an agent accesses resources that their registered profile would not permit.
- **Tabletop Exercise Schedule:** Conduct an identity-focused tabletop exercise at least once per year, including representatives from IAM, security operations, compliance, legal, and the product teams

responsible for each deployed agent. Use the exercise to identify decision-making gaps and update your incident response playbooks with pre-agreed protocols for each uncovered decision point.

- **Risk Mitigation Prioritization Matrix:** Maintain a living risk mitigation prioritization matrix that ranks your organization's identity risk exposures by likelihood and impact. Review the matrix quarterly and ensure that your top three mitigations are actively resourced with owners, timelines, and completion criteria. Do not let the matrix become a compliance artifact—use it to drive decisions.
- **Agent API Endpoint Authorization Controls:** Ensure that every registered agent's API endpoint implements caller authorization controls that restrict incoming calls to an explicitly approved list of authorized callers. Review and update the authorized caller list at every integration change, and alert on any call from an unauthorized source.
- **Incident Response Playbook for Agent Identity Events:** Develop and maintain a specific incident response playbook for AI agent identity events that pre-resolves the contain-versus-preserve dilemma, defines the regulatory notification threshold, establishes the rapid attribution process for unknown agents, and identifies the data stores that each agent has access to so that scope assessment can begin immediately.

6.6 Takeaway

Identity risks in AI systems are not extensions of traditional IT security risks. They are a new category of operational and compliance exposure that requires new detection capabilities, response protocols, and governance structures to be managed effectively. Traditional security tools were built to detect human behavioral anomalies. AI agents do not behave like humans, and the anomalies that indicate agent identity failures—quiet after-hours access that matches the agent's normal volume thresholds, permission inheritance that accumulates through automated integration workflows, shadow agents built by well-intentioned teams under delivery pressure—are precisely the kinds of signals that traditional controls are least equipped to surface. Managers who deploy AI agents without building agent-specific identity governance into their security and compliance programs are operating with a structural blind spot.

The six risk categories explored in this chapter—agent impersonation, privilege escalation, cross-agent coordination risks, identity drift, shadow agents, and identity spoofing—are not an exhaustive taxonomy of everything that can go wrong. They are the risks already manifesting in production environments across industries, and they are the risks for which governance responses are most clearly defined and most actionable. Each category has a detection approach, a mitigation strategy, and a governance process that managers can implement with existing security and compliance infrastructure if they understand what they are looking for. The challenge is not primarily technical. It is organizational: building a shared understanding among IAM, security, compliance, and product teams of what agent identity risk looks like, who is responsible for which part of the response, and what the decision protocols are when an incident occurs.

Managers who take three actions, mapping their current AI agent inventory against all six risk categories, running an annual identity-focused tabletop exercise with cross-functional participation, and maintaining a living risk mitigation prioritization matrix—will find that their organizations become measurably more resilient to agent identity incidents over time. The benefits agency in this chapter's opening scenario did not fail because its technology was inadequate. It failed because no one had designed a governance process to answer the question that the incident ultimately exposed: how do we know that the agent accessing our systems right now is the agent it claims to be? That question has a tractable answer. Building the organizational capability to ask and answer it, systematically and continuously, is the work that defines responsible AI management in the agentic era.

7 Monitor, Auditing for Agent Identities

7.1 Scenario: Audit That Nobody Could Answer

The call came on a Tuesday morning in mid-October. A pharmaceutical company's compliance director received formal notice that the regional regulatory body was initiating a structured audit of the AI-assisted drug-interaction screening system the company had deployed eighteen months earlier. The auditors wanted a complete trace: every decision the agent made, every record it flagged, every case it escalated or cleared, and the full context surrounding each action over the prior twelve months. The compliance director assumed the request was routine. She forwarded it to the IT operations team and expected a response by the end of the week.

What the operations team found instead triggered a three-week scramble. The agent's action logs existed — timestamps, database query identifiers, status codes, output classifications. What the logs did not contain was any identity context. There was no record of which version of the agent had acted on which case. There was no indication of which delegation scope was active at the time of each decision, which human had authorized the current permission set, or whether a specific action had occurred under an elevated access grant that had since been revoked. Two software updates had been deployed to the agent during the audit period. Nobody had recorded which agent identity version corresponded to which time window. The logs told the team what had happened. They could not explain who — meaning which configured, authorized instance of the agent — had done it, or why that instance had the authority to act.

The audit stalled for nine weeks. The compliance team hired an external forensic firm to reconstruct identity context from version control timestamps, deployment records, and change management tickets — none of which had been designed to link to operational logs. The reconstruction was partial at best. The regulatory body accepted the company's interim compliance assurance but imposed enhanced monitoring requirements and a corrective action plan. The direct cost of the forensic engagement exceeded $400,000. The indirect costs — delayed product approvals, reputational exposure to the regulatory body, and the eighteen months of remediation work that followed — were substantially higher. All of it traced back to a single gap: the monitoring system logged agent actions but not agent identity.

7.2 Why It Matters

7.2.1 Traditional Audit Logs Are Not Enough

Traditional IT audit logs were designed for systems that operate under a simple identity model: a human user authenticates with credentials, performs an action, and the log records the user account, the timestamp, the target system, and the action code. That model works reasonably well when the acting entity is a person whose identity is stable, whose authority is defined by a role that changes slowly, and whose decisions can be reconstructed by asking the person what they intended. Agentic AI systems break every one of those assumptions. An agent may act thousands of times per day. Its identity — meaning the combination of its model version, permission scope, delegation grants, and policy constraints — can change with each deployment update or new task delegation. The entity acting at 9:00 AM may be meaningfully different from the entity acting at 3:00 PM, even if both share the same system name.

A traditional audit log that captures timestamps, system name, action code, and IP address produces a record that indicates an action occurred. It does not tell you which configured, authorized version of the agent took that action. It does not tell you whether the agent's permission scope at the time of the action was appropriate for the action it took. It does not tell you which human or which upstream orchestrator had authorized the delegation chain active at that moment. When a regulator, internal auditor, or security team asks, "Who decided to do this, and were they authorized?" a traditional log answers the first part of the question with a system name. It cannot answer the second part at all. For agentic systems operating in regulated environments, that gap is not a technical inconvenience. It is a compliance failure waiting to be discovered.

The operational consequence is that audit readiness becomes retroactive reconstruction rather than proactive documentation. Teams spend weeks reassembling context from deployment records, change tickets, and access control snapshots that were never designed to be queried together. The cost of that reconstruction — in time, in consulting fees, in regulatory exposure — is almost always higher than the cost of building identity-enriched logging from the start.

Managers who understand this dynamic build the logging infrastructure before the audit request arrives, not after. They treat identity context in logs as a first-class governance artifact, not an optional field to be added later.

Diagram 7.1 – Traditional Audit Log vs. Identity-Enriched Agent Audit Log

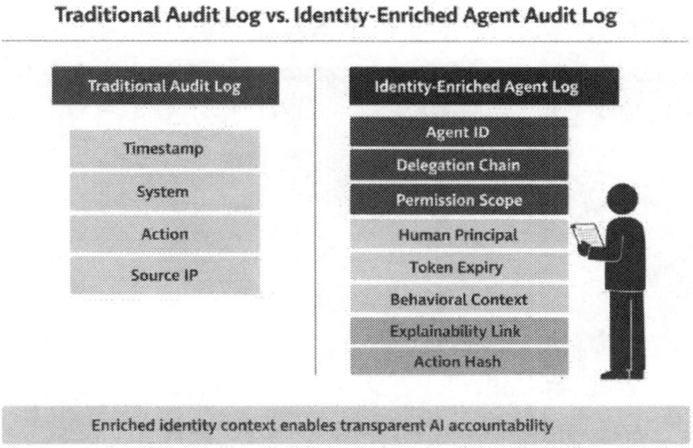

7.2.2 The Explainability Imperative

Regulators and internal governance bodies are increasingly explicit about their expectation that AI systems be explainable. The demand is not primarily about the technical interpretability of model weights or attention layers — that is a research problem. The governance demand is operational: when an AI system makes a consequential decision, the organization responsible for that system must be able to explain, in human-readable terms, what decision was made, what information it was based on, who authorized the system to make that type of decision, and whether the decision fell within the scope of that authorization. For agentic systems, that explanation is impossible to construct without an identity context. A system that cannot tell you which version of an agent acted, under what delegation, with what permission scope, has already failed the explainability requirement before the first question is asked.

Identity is a foundational pillar of AI explainability, not a secondary concern. When an auditor asks why the screening agent cleared a particular drug-interaction flag, the answer requires four components working together: the model's reasoning process (model transparency), the data the model evaluated (data lineage), the scope and authority under which the agent acted (identity and authorization), and the specific logic applied to reach the output (decision rationale). Remove any one of those four components, and the explanation is incomplete. Remove identity and authorization entirely, and the explanation cannot address the governance question at all — it can only describe what happened technically, not whether it was sanctioned. Managers who treat explainability as a model problem rather than an identity problem will build systems that satisfy data scientists but fail auditors.

The practical implication is that explainability infrastructure must be designed alongside identity infrastructure, not layered on top of it after deployment. The question regulators ask: Was this agent authorized to take this action? Who approved that authorization? Was the scope of the action consistent with the approved permission set? Was human oversight invoked at the appropriate thresholds? — are identity questions. The monitoring and logging systems that produce audit-ready answers to those questions must capture identity data as a native part of every logged event. Retrofitting identity context into an existing logging infrastructure is expensive and often incomplete. Building it in from the start is a design decision that managers must drive, because engineers under delivery pressure will not prioritize it without explicit direction.

Diagram 7.2 – Identity as a Pillar of AI Explainability

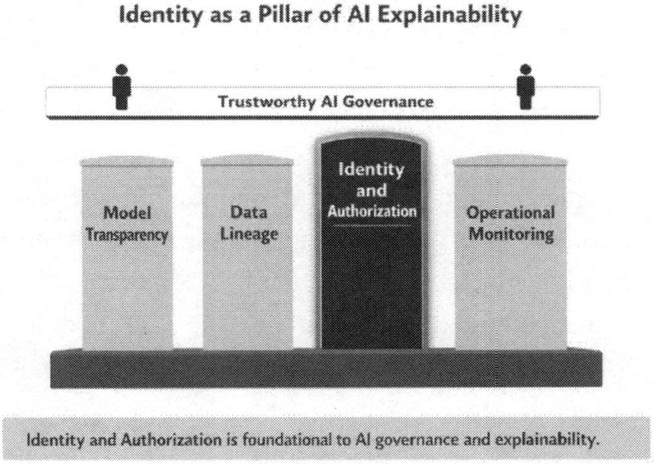

7.3 Framework: Build Identity-Aware Monitor/Audit

7.3.1 Identity-Linked Logging

Every agent action that could affect a business outcome, a data record, or a downstream system must generate a log entry that contains a minimum set of identity fields. The agent identifier must be present — not just the system name, but the specific versioned identity of the agent at the time of the action, including the model version, the configuration snapshot hash, and the deployment identifier. The delegation chain must be recorded: which orchestrator or upstream principal granted this agent authority for this task, and through what chain of delegated permissions did that authority flow. If the agent was granted elevated access for a specific workflow, the log must record the scope of that elevation and the policy version under which it was granted.

Human authorization context must be captured explicitly. For every permission scope active at the time of an action, the log must record which human principal approved that scope, when the approval was granted, and when it expires or was last reviewed. This is not the same as recording the human who initiated the current session — it is a record of the governance chain that authorized the agent's current capability set. That distinction matters in regulatory contexts where auditors are not just asking who used the system, but who is accountable for the system's authorization to act. The log entry is not just an operational record — it is a governance artifact that establishes accountability.

Action outcome and oversight status complete the identity-linked log entry. The outcome — what the agent produced, at what confidence level, and with what effect on downstream systems — provides auditors with the factual basis to evaluate whether the action was appropriate. The oversight status — whether the action was autonomous, whether a human reviewed it before execution, and whether it triggered an alert — provides auditors with the governance basis to evaluate whether the oversight framework functioned as designed. A log entry that contains all of these fields does not just record what happened. It records who acted, under what authority, with what accountability structure in place, and what resulted. That is the audit-ready standard that identity-aware logging must meet.

7.3.2Context-Aware Audit Trails

An audit trail is not the same as a collection of log entries. A log entry records a single event. An audit trail reconstructs the complete governance context of a sequence of events: what the agent was doing, why it had authority to do it, how that authority was established, and whether the human oversight framework operated correctly throughout. Building audit trails for agentic systems requires an enrichment layer between the raw log events and the audit store. This pipeline takes individual log entries and attaches the contextual data needed to reconstruct the governance context. That context includes the delegation chain as it existed at the time of each event, the policy version active at that time, the human approval records associated with the active permission scope, and the access control state of every system the agent interacted with.

The enrichment pipeline must operate in near-real time to be useful for triggered investigations. If an anomaly is detected — an agent accessing a system outside its documented permission scope, a delegation chain that does not match the authorized topology, an action taken at a time when the authorizing human approval had expired — the investigation team needs enriched context immediately, not after a batch enrichment process completes twelve hours later. Building enrichment as a streaming capability, rather than a batch process, is a design choice with real operational consequences. It is also a cost decision: streaming enrichment infrastructure costs more to operate than batch processing, but the operational value of immediate context availability during an active incident justifies the investment for regulated workloads.

The audit store itself must be designed for governance queries, not just operational retrieval. Operational log stores are optimized for time-range queries: give me all events from this system between these two timestamps. Governance audit stores need to support identity-centric queries: give me all actions taken by agent version 2.3.1 under delegation from orchestrator-001, within permission scope X, during the period when policy version 4.2 was active, and show me any actions that exceeded the approved scope. Those query patterns require indexing on identity fields, policy version fields, and delegation chain fields — not just on timestamps and system names. Designing the audit store schema for governance queries is a prerequisite for building a monitoring system that can answer auditor questions.

Diagram 7.3 – Context-Aware Audit Trail Architecture

Context-Aware Audit Trail Architecture

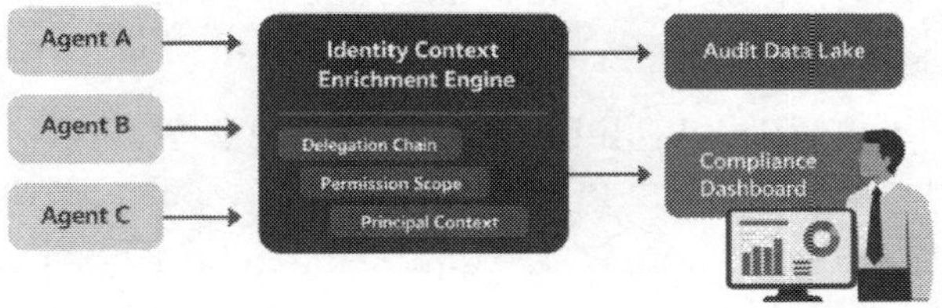

Identity enrichment transforms raw events into audit-ready evidence.

7.3.3 Dashboards and Alerts

Monitoring dashboards for agent identity serve a different purpose than operational performance dashboards. A performance dashboard shows you whether the system is running — latency, throughput, error rates, and availability. An identity monitoring dashboard shows you whether the system is running correctly within its authorized boundaries — whether agents are using permissions as expected, whether delegation chains are intact and authorized, whether credentials are healthy and approaching refresh thresholds, and whether access patterns are consistent with the established baseline. These two types of information serve different audiences: operations teams need performance dashboards, while governance teams need identity-monitoring dashboards. In many organizations, neither audience is fully served because the monitoring infrastructure conflates operational metrics with identity governance signals.

A well-designed identity monitoring dashboard organizes information into four governance-relevant panels.

The permission usage heatmap shows which permissions are being exercised by which agents over time, making it immediately visible when an agent begins using a permission it has historically not needed — a pattern that may indicate misconfiguration, scope creep, or compromise. The delegation chain violations panel surfaces any instance where an agent's active delegation chain differs from the authorized topology: a sub-agent receiving authority from an orchestrator not listed in the approved delegation registry, or a delegation depth exceeding the policy-defined maximum. The credential health panel tracks the age and renewal status of every agent credential, alerting before credentials expire and flagging credentials that have not been rotated within the policy-defined window. The access pattern anomalies panel applies baseline behavioral analysis to detect deviations: an agent accessing systems at unusual hours, querying record volumes outside its historical range, or initiating outbound connections to systems not in its documented integration list.

Alert thresholds for identity anomalies require calibration that operational teams rarely do automatically. A credential within thirty days of expiration should generate a low-priority alert to the agent's owning team. A credential that is within seven days of expiration should trigger a high-priority alert to the governance lead. A delegation chain violation should generate an immediate alert to both the security team and the agent's owning manager, with automatic suspension of the non-conforming delegation pending review. An access pattern deviation that exceeds two standard deviations from the established baseline should trigger an investigation workflow, not just a dashboard notification. Building these alert escalation paths into the monitoring design — rather than leaving them to individual team discretion — is the difference between a monitoring system that surfaces issues and one that merely records them.

Diagram 7.4 – Identity Monitoring Dashboard Layout

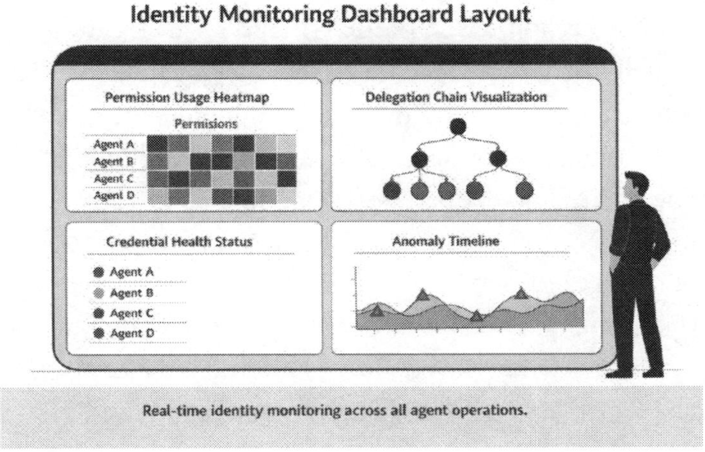

7.3.4 Audit Workflows for Agent Identity

Effective agent governance requires two distinct types of audit workflows: scheduled reviews and triggered investigations. Scheduled reviews operate on a calendar cadence — typically quarterly for routine access and permission scope reviews, monthly for credential health audits, and annually for full delegation chain and policy alignment assessments. Scheduled reviews are the governance equivalent of a recurring maintenance check. They exist not because something has gone wrong but because governance discipline requires periodic verification that everything is operating within its authorized parameters. A manager who schedules and executes these reviews consistently builds an institutional record that demonstrates proactive governance — the kind of record that turns a regulatory audit from a crisis into a formality.

Triggered investigations are the response to anomalies surfaced by the monitoring system. A delegation chain violation, a credential anomaly, an access pattern deviation, or a permission usage outlier should each trigger a defined investigation workflow that assigns ownership, establishes a response timeline, and produces a documented finding. The workflow does not need to be complex. A simple three-step process of detection, investigation, and disposition is sufficient — but it must be documented, consistently followed, and tied to the monitoring system's alert output. An alert that generates a dashboard notification but does not trigger an assigned workflow is not a governance control. It is a record of a problem that nobody was required to address.

Continuous compliance monitoring sits between scheduled reviews and triggered investigations. It is the ongoing baseline verification that agent identities, delegation chains, permission scopes, and credential states are consistent with the current policy set — not just at audit time but every day. Building continuous compliance monitoring into the monitoring infrastructure means that the governance team has a real-time view of policy conformance, not a point-in-time snapshot that may be stale within hours of a deployment update. Managers who implement continuous compliance monitoring do not scramble to answer audit questions because the audit-ready answer is always current. The monitoring system is not producing a record of what happened — it maintains a live governance posture, always ready for examination.

7.3.5 Bridging Monitoring and Explainability

Monitoring data is the raw material for explainability. When a regulator, internal auditor, or compliance officer asks, "Why did this agent take this action?" the answer must be drawn from the monitoring system's enriched audit trail. The question is not just a request for a description of what the agent did — it is a request for a complete governance account: which agent, in which configured state, under which delegation, with which permission scope, authorized by whom, operating under which policy, took which action, with what outcome, and whether the appropriate oversight mechanisms were engaged. A monitoring system designed without explainability in mind will have pieces of that answer scattered across multiple systems, requiring manual reconstruction. A monitoring system designed with explainability as a first-class requirement will be able to generate that complete governance account automatically, on demand, in response to a query.

The bridge between monitoring and explainability is a deliberate design choice, not an emergent property of good logging. Managers must specify, before the monitoring system is built, what questions it must answer and in what form those answers must take. The regulatory explainability question — fully answered, with identity context, delegation chain, policy reference, and oversight status — should be treated as a design requirement, not a post-deployment nice-to-have. Operational clarity about the agent's authorized state at every moment is what makes governance tangible. It converts the abstract commitment to responsible AI into a concrete, queryable, auditor-ready record. That conversion is the manager's job to design and the monitoring system's job to execute.

Diagram 7.5 – From Monitoring Event to Explainability Answer

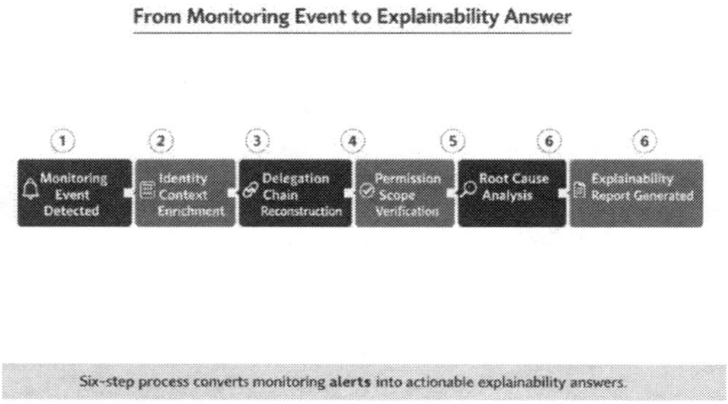

7.4 Example: Designing Audit-Ready Monitor Sys.

A large insurance company operating in multiple regulated jurisdictions had deployed twelve AI agents across its claims processing, fraud detection, and policy underwriting workflows over three years. Each agent had been deployed by a different team, using whatever logging standard that team had adopted at the time. When the company's enterprise risk committee conducted an internal governance review, it found that no two agents produced logs in compatible formats, none of the logs contained consistent identity fields, and the company could not produce a unified governance view of its agent portfolio at any level of abstraction. A regulatory inquiry shortly after the internal review confirmed the risk: the company received a request for evidence of governance controls over its AI underwriting agent. It was unable to respond with documentation that satisfied the regulator's identity-context requirements. The remediation project that followed was framed explicitly as building an audit-ready monitoring system. This unified infrastructure would cover all twelve agents and be designed from the start to satisfy regulatory explainability requirements.

The project was structured in four phases over sixteen weeks. Weeks one through four focused on logging schema redesign. The team — drawing on security, compliance, engineering, and operations — defined a canonical identity-enriched log format applicable to all agents, regardless of the underlying technology stack. The schema specified twelve required identity fields: agent identifier, model version, configuration snapshot hash, deployment identifier, delegation chain array, authorizing orchestrator identity, permission scope list, active policy version, policy approval date, human authorization record, action identifier, and oversight status. Every team operating an agent was required to instrument its agent to emit logs conforming to this schema. The compliance lead served as the schema owner, with authority to reject agent deployments that did not meet the logging standard. The manager's role in this phase was primarily governance: establishing the schema as a mandatory standard and resolving the team conflicts that arose when existing agents required re-instrumentation.

Weeks five through eight deployed the enrichment engine — the pipeline that ingested raw agent logs and attached the contextual data needed to support governance queries. The enrichment engine, integrated with the company's identity governance platform, pulls delegation chain records in real time; the policy management system attaches the current and historical policy versions associated with each permission scope; and the human authorization registry links each active scope to its approving human principle. The engineering lead managed technical integration, while the security lead reviewed the enrichment pipeline's access controls to ensure that the pipeline itself did not become an identity vulnerability — a system with access to the complete delegation chain and policy registry of every agent in the portfolio is a high-value target and must be governed accordingly. By the end of week eight, the enrichment engine was processing live log events from all twelve agents and producing governance-ready audit records within seconds of each event.

Weeks nine through twelve built out the four monitoring dashboard panels — permission usage heatmap, delegation chain violations, credential health status, and access pattern anomalies — and configured the alert escalation paths. The operations manager drove the dashboard build in partnership with the compliance team, who specified the alert thresholds based on the company's regulatory commitments and internal risk tolerance. Weeks thirteen through sixteen conducted audit workflow testing: the team ran three simulated regulatory audit scenarios, each requiring a complete governance account of a specific agent decision within a defined time window. The first simulation took eleven hours to complete, still requiring significant manual work. The second took four hours. The third, after workflow refinements, took forty-seven minutes and produced a complete, formatted explainability report with full identity context, delegation chain, policy reference, and oversight status. The company's compliance director presented the results to the enterprise risk committee as evidence that the audit-ready monitoring system was operational, and the governance gap had been closed.

Diagram 7.6 – Audit-Ready Monitoring Implementation Roadmap

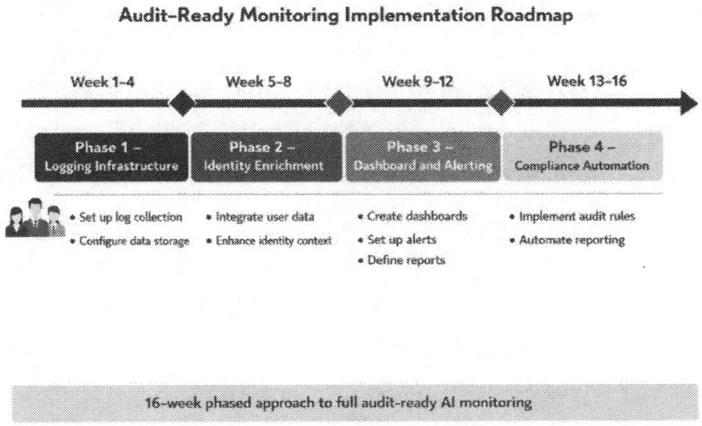

7.5 Manager's Checklist

- **Identity-enriched logging schema:** Define and mandate a canonical log format that includes agent identifier, model version, deployment snapshot, delegation chain, permission scope, active policy version, human authorization record, and oversight status for every agent action.
- **Audit trail enrichment pipeline:** Deploy a real-time enrichment engine that attaches governance context to raw log events, integrating with the

identity governance platform, policy management system, and human authorization registry.

- **Identity monitoring dashboard:** Build a four-panel governance dashboard covering permission usage patterns, delegation chain conformance, credential health status, and access pattern baselines — distinct from operational performance dashboards.
- **Alert thresholds and escalation paths:** Configure calibrated alert thresholds for credential age, delegation chain violations, permission scope anomalies, and access pattern deviations, with defined escalation paths that assign ownership and require documented disposition.
- **Audit workflow documentation:** Establish documented workflows for both scheduled reviews (quarterly, monthly, annual) and triggered investigations, with assigned owners, response timelines, and required outputs for each workflow type.
- **Explainability response template:** Define the standard format for regulatory and internal audit explainability responses, specifying the identity fields, delegation context, policy references, and oversight status documentation required in every answer.
- **Regulatory alignment verification:** Confirm with legal and compliance that the logging schema, enrichment pipeline outputs, and audit workflow documentation satisfy the specific regulatory frameworks governing each deployed agent,

including any sector-specific requirements for AI decision traceability.

- **Quarterly audit readiness review:** Conduct a structured quarterly review of audit readiness — simulating a regulatory inquiry, measuring response time, and testing the completeness of the explainability report produced — to maintain governance discipline and surface degradation before a real audit request arrives.
- **Logging standard enforcement gate:** Establish the identity-enriched logging schema as a mandatory deployment gate for all new agents, with the compliance or governance lead holding authority to block deployments that do not meet the standard.
- **Audit store schema design for governance queries:** Ensure the audit store indexes on identity fields, policy version, delegation chain, and human authorization records — not just on timestamps — so governance queries can be executed directly without manual reconstruction.

7.6 Takeaway

Monitoring and auditing agent identities is not an overhead added to an already-complex deployment. It is the operational mechanism that converts the commitment to responsible AI into something tangible, verifiable, and defensible. Every governance principle — least privilege, delegation discipline, human oversight at appropriate thresholds, and policy-aligned access control — is only as strong as the organization's ability to demonstrate that it is being upheld. That demonstration requires monitoring data.

It requires enriched audit trails. It requires dashboards that surface anomalies before they become incidents, and audit workflows that produce complete governance accounts on demand. Managers who invest in this infrastructure are not building compliance theaters. They are building an operational foundation that makes every other governance commitment real.

The explainability dimension of that investment deserves particular emphasis. The question regulators, auditors, and governance boards are increasingly asking of AI systems is not just whether the system performed correctly — it is whether the organization can prove it. Proof requires an identity context. It requires a complete account of who acted, under what authority, with what scope, authorized by whom, and with what oversight. A system that produces correct outputs but cannot answer those questions is not a governed system — it is a trusted-but-unverified capability, and trust without verification is a liability in any regulated environment. Building the monitoring and audit infrastructure to answer those questions completely and efficiently is one of the highest-leverage investments a manager can make in the governance of agentic AI.

The managers who do this work well share a common characteristic: they treat operational clarity as a strategic asset rather than an administrative burden. They design monitoring systems that make invisible governance visible. They build audit workflows that produce answers in minutes rather than weeks. They define explainability response templates before the first audit request arrives, not after. They conduct quarterly readiness reviews that keep the governance posture current rather than letting it drift between external audits. The result is an organization that can confidently deploy agentic AI in complex, regulated environments because its governance infrastructure is as robust as its technical infrastructure. That confidence is not incidental — it is the mission-aligned outcome of building responsible AI capabilities the right way, from the start.

8 Integrate Identity into AI SDLC and Pipelines

8.1 Scenario: The Deployment That Bypassed Identity

Meridian Financial Services ran a tightly scheduled machine learning operation. Its fraud detection platform had been in production for two years, and the team had developed a fast-track CI/CD pathway specifically for model updates—changes to weights, thresholds, and scoring logic that did not alter the agent's behavioral architecture or integration surface. This fast-track pathway bypassed the identity review gate, which was designed for agent deployments that carry new system access requests, credential configurations, or permission scope changes. It was a reasonable accommodation, given that pure model updates carried little identity risk. The problem began when a model update was not actually a pure model update.

A junior engineer on the ML team submitted a deployment manifest classifying the new fraud detection build as a model update. The classification was partly an oversight and partly a misunderstanding of the team's own taxonomy: the building included a new data enrichment module that reached into a transaction archive the previous version had never touched. The deployment manifest inherited permission scopes from the previous deployment's configuration file, which had been written for a broader integration prototype and never cleaned up. Because the identity review gate was bypassed, no one validated the permission scope against the functional requirements of the current version. This build passed automated testing, cleared the reduced fast-track pipeline, and went to production within four hours of submission. Two weeks later, an anomaly-detection rule flagged an access pattern: the fraud agent was reading transaction records from a data-warehouse segment outside its authorized boundary. The agent was not malfunctioning. It was functioning exactly as its inherited permissions allowed.

The incident triggered a mandatory internal review. The root cause was unambiguous: a deployment classification decision made by a developer had bypassed a governance gate designed to catch exactly this kind of permission creep. The organizational response was equally clear. The team retired the model-versus-agent taxonomy that had created the gap in the first place, replacing it with a single classification standard based on permission scope change rather than deployment type. Any build that touched credential configuration, permission manifests, or integration surface—regardless of what else it changed—was required to pass the identity review gate. The security team mandated a full audit of the deployment manifest library to identify other configurations carrying over-provisioned permissions from previous builds. Twelve manifests were flagged. Three were in active production. The incident cost the team six weeks of remediation work and generated a compliance finding that required executive sign-off to close. The fast-track pathway that was meant to accelerate delivery had instead produced the slowest quarter the ML team had experienced in three years.

8.2 Why It Matters

8.2.1 Identity Is a Development Concern, Not Just an Operations Concern

The instinct to treat identity as an operation's concern is understandable. Infrastructure teams provision credentials. Security teams configure access policies. Compliance teams review audit logs. These are all operational activities, and they happen after the code is written. But the decisions that determine what identity an AI agent will carry—what systems it will touch, what permissions it will request, what delegation model it will use—are made at design time. By the time the agent reaches operations, its identity posture is largely locked in. Retrofitting identity controls onto a deployed agent is expensive, disruptive, and often incomplete. The manager who treats identity as a design-day commitment builds agents that are responsible by default. The manager who defers identity to operations builds agents that require emergency remediation.

Every stage of the AI development lifecycle carries identity implications. At design, the team decides what the agent needs access to and begins sketching its permission profile. During development, engineers integrate credential management into the agent's runtime and implement token request patterns. At testing, the team validates that the agent respects permission boundaries and handles denied requests appropriately. At deployment, an identity review gate confirms that the permission manifest matches the agent's functional design. During monitoring, the team watches for access anomalies that suggest permission drift or misuse. At retirement, the team ensures that credentials are revoked and audit records are preserved. Skipping any of these touchpoints creates a gap that adversaries, configuration errors, or simple misclassification can exploit.

Managers who understand this end-to-end picture can build processes that distribute identity responsibility across the entire team, not just the security or operations function. Developers own credential integration standards. Test engineers own permission boundary test cases. Release managers own identity review gates. Security architects own red-team scenarios. This distribution is not a burden—it is the mechanism that makes identity governance sustainable at the pace of delivery. A security review that happens once at the end of a cycle is a bottleneck. Identity checks embedded at every stage are part of the workflow, and they cost a fraction of the remediation bill when issues surface in production.

Diagram 8.1 – Identity Touchpoints Across the AI SDLC

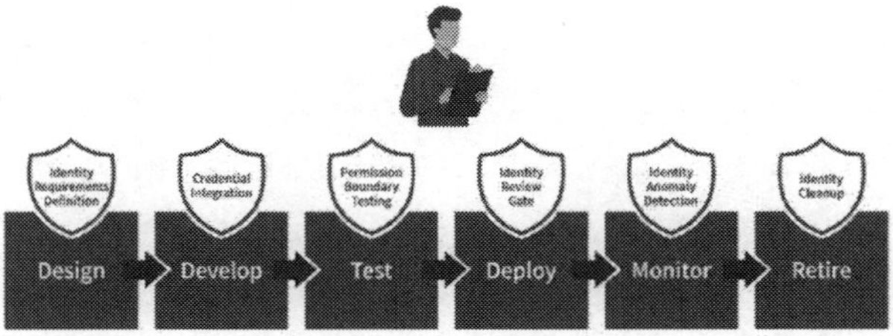

8.2.2 The Cost of Late Integration

When identity is treated as a post-deployment concern, the consequences are severe. A permission issue identified at the design stage costs an hour of conversation and a revised architecture sketch. The same issue was discovered during testing, costing days of rework: developers must refactor credential integration, test engineers must rerun boundary validation suites, and security architects must re-review the updated configuration. Discovered in production, the same issue costs weeks: incident response, compliance investigation, executive notification, customer impact assessment, remediation across all affected environments, and a post-incident review that generates its own action list. The relative cost curve is not linear. Organizations that have studied their incident histories consistently find that production identity issues cost 30 to 50 times as much as design-stage prevention would have cost.

The organizational damage extends beyond the remediation bill. Compliance findings related to unauthorized data access carry regulatory exposure in virtually every sector that deploys AI for consequential decisions—financial services, healthcare, federal contracting, and critical infrastructure. A single production identity incident can stall an entire AI program while the compliance team works through its findings. Release schedules freeze. Product roadmaps shift. Engineering bandwidth allocated to new capabilities is redirected to audit responses and control attestations. The team that was building the future spends a quarter explaining the past.

Managers who internalize the cost curve of late identity integration make better prioritization decisions. When a developer pushes back on adding permission boundary test cases because it adds a day to the sprint, the manager who understands the cost curve knows that a day of prevention is worth weeks of remediation. When a product leader suggests skipping the identity review gate to hit a launch date, the manager who understands the cost curve can quantify the risk in terms that the product leader's P&L will recognize. Identity governance is not a tax on delivery speed. It is insurance against the events that stop delivery entirely.

Diagram 8.2 – Cost of Identity Integration by SDLC Stage

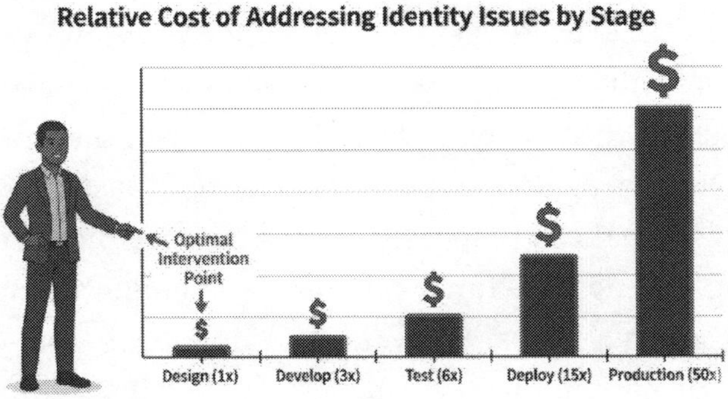

8.3 Framework: Identity Integration Points

8.3.1 Agent Design and Identity Requirements

Identity requirements belong in the agent's design document, alongside functional, data, and performance requirements. Before the first line of code is written, the design team must answer a set of foundational questions:

What systems will this agent access? What data will it read, write, or modify? What permissions does it need to perform its defined function, and nothing more? Which delegation model will it use when acting on behalf of human users or other agents? What audit trail must it produce to satisfy governance and compliance requirements? These answers define the agent's identity profile—the complete specification of who the agent is, what it is authorized to do, and how it will account for its actions.

The identity profile established at design time serves as the reference document for every subsequent identity decision throughout the lifecycle. During development, engineers use it to configure credential integration. During testing, test engineers use it to define permission boundary test cases. During the CI/CD identity review gate, the reviewer validates that the deployment manifest matches the agent's intended design. During monitoring, the operations team uses it to define the normal access baseline against which anomalies are measured. A well-constructed identity profile at design time eliminates ambiguity at every downstream stage. Conversely, an agent that enters development without a defined identity profile will acquire one through accident, through whatever permission happen to be available, inherited, or convenient.

Managers should require identity profiles as a gate condition for moving from design to development. This does not mean the identity profile must be complete and final before development begins—requirements evolve. The profile should be a living document, version-controlled alongside the agent's codebase. But the initial identity profile must exist, be reviewed by a security architect or identity governance lead, and reflect a genuine analysis of the agent's access requirements. An identity profile that says 'TBD' or lists broad system access without a functional justification is not a profile—it is an open permission request that will cause downstream problems.

8.3.2 Development: Credential Integration and Secure Coding

Credential integration is where identity requirements become code. The foundational principle is simple: AI agents must never carry hardcoded credentials. Hardcoded secrets—API keys, service account passwords, database connection strings embedded in source code or configuration files—are a persistent and well-understood vulnerability. They appear in version control history, in building logs, in container images, and in any environment where the code runs. For AI agents that operate with elevated permissions and handle sensitive data, hardcoded credentials provide a direct path from code repository access to unauthorized data access. The development standard must be unambiguous: credentials are retrieved at runtime from a credential vault, never embedded in code or configuration.

The implementation pattern for secure credential integration follows a consistent model. The agent's runtime includes a credential vault SDK that is configured to request scoped, short-lived tokens at the start of each task execution. The token carries only the permissions required for the specific task the agent is performing—not a broad set of permissions that cover every possible task the agent might ever perform. When the task is complete, the agent releases the token, and the vault records the credential lifecycle for audit purposes. This pattern—request, use, release limits the risk of long-lived credentials being intercepted or misused between task executions. It also creates a natural audit trail: every credential request is logged, every token is scoped, and every release is recorded.

Developers must also build and test credential rotation tolerance. In any mature enterprise environment, credentials rotate on a schedule—service account passwords expire, API keys are cycled, and certificates are renewed. An AI agent that cannot handle credential rotation gracefully will fail silently or fail loudly at the worst possible moment. The development standard should require that every agent implementation be tested against simulated credential rotation: the vault issues a new token mid-task, and the agent must request the new token and continue operation without interruption. Credential rotation testing is not an edge case—it is a baseline requirement for any agent that will operate in production for more than a few months.

Diagram 8.3 – Secure Credential Integration Pattern for Agent Development

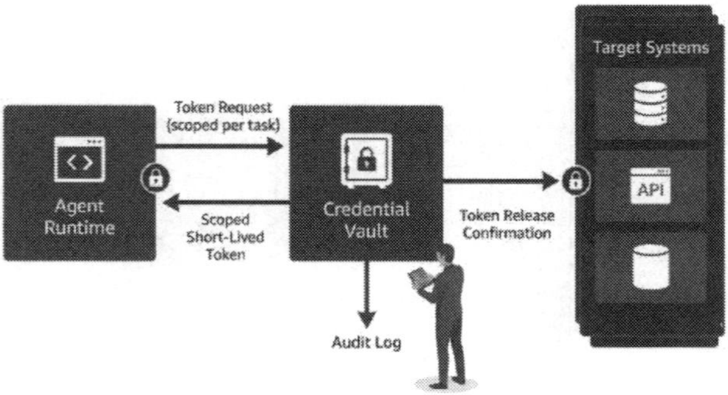

8.3.3 Testing: Permission Boundary Validation

Testing an AI agent's identity behavior requires a deliberate test suite organized around its defined permission boundaries. The foundational question is: can the agent access only what its identity profile says it should access? Test engineers must construct scenarios that probe both sides of every permission boundary—confirming that authorized access succeeds and that unauthorized access fails. An agent that successfully retrieves data has passed one half of the test. An agent that correctly handles a denial when it attempts to access data outside its scope has passed the other half. Both results are required to pass the permission boundary test. An agent that handles denied requests by crashing, generating unhandled exceptions, or silently skipping required steps is not production-ready, regardless of whether its authorized access functions correctly.

Delegation boundary testing is a distinct and often overlooked category. When an agent acts on behalf of a human user, it should carry no more authority than that user would carry in the same transaction. Test engineers must verify that the agent's delegation chain is correctly implemented: when User A, who has read-only access to a data set, delegates a task to the agent, the agent must not execute that task with written permissions that User A does not possess. This sounds obvious in principle, but is frequently misconfigured in practice. Delegation chains that are implemented in code without explicit boundary enforcement tend to collapse to the highest-permission principal in the chain. Testing must specifically target these collapse scenarios and confirm that the agent enforces the least privilege constraint at every delegation link.

Permission boundary test results must be treated as release gates, not advisory findings. An agent that fails permission boundary tests is not ready for deployment, regardless of how well it performs on functional tests. The test report must document every boundary scenario tested, every denial handled, and every delegation chain validated. This documentation becomes part of the agent's identity record and supports the CI/CD identity review gate that follows. Managers should require that the test suite be version-controlled alongside the agent's codebase, updated whenever the agent's identity profile changes, and re-executed as part of every build that touches credential configuration or permission scope.

8.3.4CI/CD: Identity Review Gates

The CI/CD pipeline is where identity governance becomes operational enforcement. An identity-aware pipeline treats permission manifests, credential configurations, and agent identity profiles as first-class artifacts that must be validated at every build, not as supporting documentation that is reviewed once and filed. The pipeline should include at minimum three automated identity validation stages: a permission manifest check that confirms the deployment manifest's declared scopes match the agent's approved identity profile; a credential configuration audit that verifies no hardcoded secrets exist in the build artifacts and that vault integration is correctly configured; and a permission scope drift check that compares the current build's declared permissions against the previous production deployment and flags any expansions for human review.

Automated gates reduce friction and eliminate the class of issues caused by human oversight under time pressure. But automated gates alone are insufficient for agent deployments. Any build that introduces a change to the agent's permission scope, delegation model, or credential configuration requires a human identity review before deployment to production. This human gate is not optional and is not subject to fast-track bypass regardless of how the deployment is classified. The reviewer—a security architect, identity governance lead, or authorized release manager—must confirm that the identity profile change is intentional, has been tested, and is justified. The review should take minutes, not days, if the upstream documentation is complete. When it takes days, it signals that upstream documentation is incomplete.

Post-deployment identity monitoring should be activated during the deployment step itself, not as a separate post-launch task. The monitoring configuration—defining what constitutes normal access behavior for this agent version, what thresholds trigger anomaly alerts, and what response procedures apply—should be included in the deployment manifest and validated at the same gate that validates permission scope. An agent that goes to production without an active monitoring configuration is an agent operating without accountability. Managers who require monitoring activation as a deployment gate condition create an environment where accountability is structural rather than aspirational.

Diagram 8.4 – Identity-Aware CI/CD Pipeline

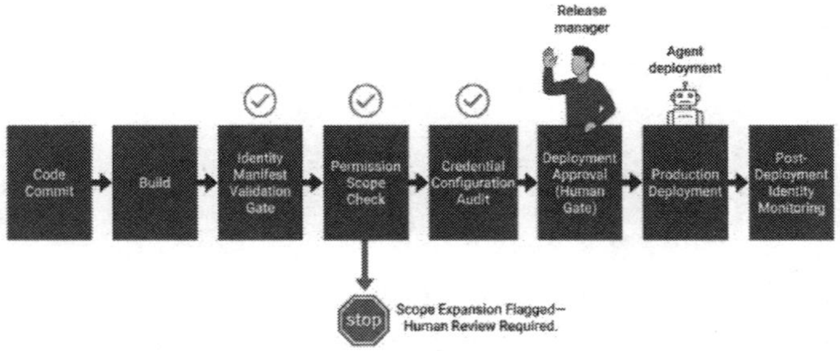

8.3.5 Red-Teaming for Identity

Red-teaming exercises for AI systems commonly focus on adversarial inputs, output manipulation, and model evasion. Identity-focused red-teaming adds a distinct and critical dimension: can the agent's identity be exploited? Red-team scenarios for identity should include attempts to manipulate the agent into requesting broader permissions than its task requires—for example, by crafting inputs that trigger fallback behaviors with elevated access; attempts to spoof the agent's identity to other systems in the integration chain; and attempts to manipulate the delegation chain so that the agent acts with the authority of a higher-permission principal than the one who initiated the task. These scenarios should be developed by security architects familiar with the agent's integration surface and run as part of any major release cycle.

The output of identity red-teaming is not a pass/fail grade, it is a library of scenarios that informs the agent's ongoing security posture and the team's understanding of its attack surface. Findings from red-team exercises should feed directly into the permission boundary test suite, the credential integration standards, and the monitoring alert configuration. A red-team finding that demonstrates a delegation chain collapse, for example, should generate a new permission boundary test case that is added to the permanent test suite and re-executed on every subsequent build. Managers should require that identity red-team findings be tracked to closure with the same rigor as any other security vulnerability, with a defined remediation timeline and a verification step confirming the fix is effective.

8.3.6Model Registries and Agent Orchestration

Model registries serve as the authoritative catalog of every AI model and agent version that an organization has built, tested, and approved for use. For registries to function as identity governance infrastructure—not just version control, every entry must include a complete identity profile alongside the technical artifacts. A registry entry that contains model weights, performance benchmarks, and deployment instructions but lacks an identity profile is incomplete from a governance perspective. The required identity metadata for every registry entry includes agent name and version, the approved identity profile with permission scope, the credential type and vault integration configuration, the delegation model, audit requirements, and the deployment approval status indicating whether the entry has cleared the human identity review gate. Without this metadata, the registry cannot answer the governance questions that matter: what is this agent authorized to do, who approved it, and has that approval been validated?

Agent orchestration framework systems that coordinate the behavior of multiple AI agents working in concert on complex tasks—introduce layered identity challenges that registries must be equipped to surface. When an orchestrator assigns a subtask to a subordinate agent, it must be able to confirm that the subordinate agent's identity profile is consistent with the orchestration design: that the subordinate has the permissions required for its assigned task, that its delegation model is compatible with the orchestrator's authority chain, and that the orchestration framework's logging infrastructure meets its audit requirements. A registry that carries complete identity metadata for every agent version makes these confirmations automated and reliable. A registry that carries only technical metadata forces the orchestration team to chase identity information from multiple sources at deployment time—creating exactly the kind of time-pressure gap that produces misclassification errors and bypassed review gates.

Diagram 8.5 – Identity-Enriched Model Registry

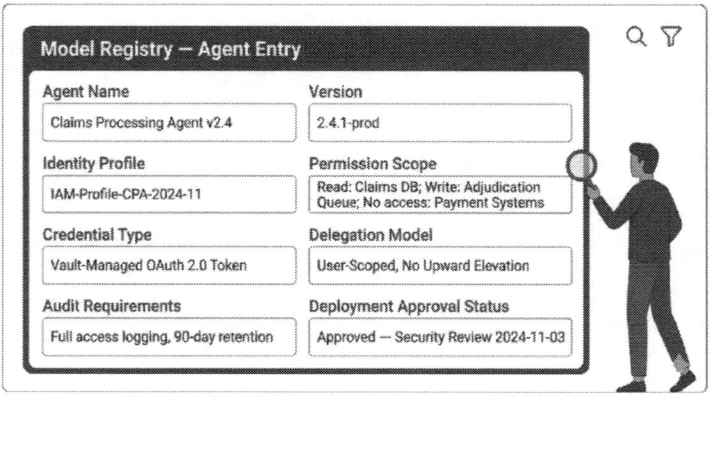

8.4 Example: Building an Identity-Integrated Pipeline

Navarro Federal Health Solutions, a mid-sized IT contractor serving federal health agencies, operated a portfolio of AI claims-processing agents that handled adjudication routing, eligibility verification, and payment recommendations for multiple agency clients. The agents processed tens of thousands of claims per day and had access to protected health information under strict federal data governance requirements. After a compliance audit identified three agents operating with over-provisioned permissions inherited from earlier development configurations, leadership authorized a twelve-week initiative to build a fully identity-integrated CI/CD pipeline. The goal was not just to fix the three flagged agents but to make the identity integration problem structurally impossible to repeat for any future deployment.

The first three weeks focused on definition and instrumentation. The security architecture team developed a standard identity manifest schema structured YAML document that every agent build would be required to produce. The schema defined required fields for permission scope, credential type, delegation model, audit requirements, and approval status. The DevOps team built an automated validation gate that parsed the identity manifest at build time and rejected builds that were missing required fields or declared permission scopes inconsistent with the agent's registered identity profile. They also built a scope drift detector that compared the incoming manifest against the previous production build and generated a human review ticket for any permission expansion. By the end of week three, the automated gates were live in the staging environment. They had already caught four builds—two with missing audit configuration, one with an undeclared credential change, and one with a permission scope that had expanded beyond its approved profile without explanation.

Weeks four through eight addressed the human workflow. The team designed a human-approval workflow for agent deployments that triggered whenever the automated gates flagged a permission-scope change, a new delegation model, or a new system integration. The workflow routed to a security architect within the first four hours of a flagged build and required a documented decision—approve, reject, or request revision—within one business day. To make this feasible at the team's release cadence, the security architects were given read-only access to the full build context: the identity manifest, the diff against the previous approved version, the permission boundary test results, and the red-team finding library for the agent. The approval decision took an average of forty minutes per build. The team had expected the human gate to create a bottleneck. Instead, it created accountability: developers began producing cleaner identity manifests because they knew a security architect would read them.

The final four weeks extended the pipeline to cover post-deployment identity monitoring. Each deployment now automatically activates a monitoring profile configuration pushed to the centralized SIEM that defines the agent's expected access baseline, the alert thresholds for anomalous behavior, and the response procedure for each alert category.

The monitoring profile was derived directly from the approved identity manifest, so the expected baseline was always aligned with what the agent was approved to do. By the end of week twelve, all seventeen active agents in the portfolio had been migrated to the new pipeline. The team measured deployment confidence through a proxy metric: the number of identity-related post-deployment corrections required per quarter. In the two quarters before the initiative, the team averaged eleven identity corrections per quarter. In the first quarter after full migration, the number dropped to one—a misconfigured audit log retention setting was caught and corrected within forty-eight hours of deployment. The pipeline had made identity governance a property of the delivery process, not a separate remediation workstream.

Diagram 8.6 – Identity-Integrated Pipeline Architecture

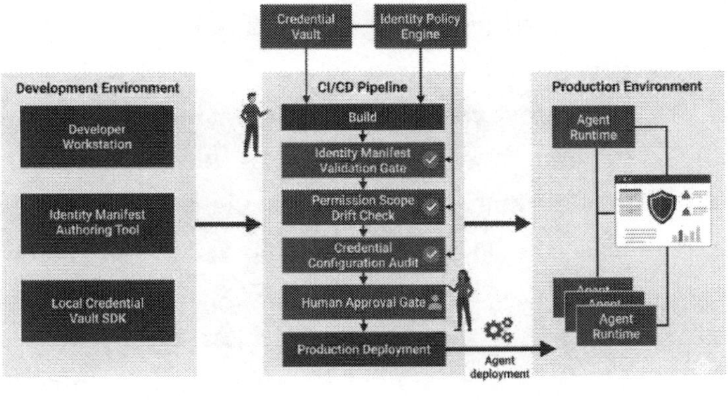

8.5 Manager's Checklist

- **Identity requirements in design documents:** Require that every agent design document include a completed identity profile covering permission scope, delegation model, credential type, and audit requirements before the project moves to development.

- **Credential integration standards:** Enforce team-wide standard prohibiting hardcoded credentials in any agent codebase or configuration artifact, and require vault-managed, short-lived token patterns for all runtime credential access.
- **Permission boundary test cases:** Require that every agent build include a permission boundary test suite covering both authorized access confirmation and unauthorized access denial handling, with delegation chain boundary scenarios treated as release gates.
- **CI/CD identity gate configuration:** Ensure the CI/CD pipeline includes automated identity manifest validation, permission scope drift detection, and credential configuration audit gates, with human review triggered by any permission scope change or new system integration.
- **Red-team scenario library for identity:** Maintain a version-controlled library of identity-focused red-team scenarios—permission elevation manipulation, identity spoofing, delegation chain collapse—updated after every major release and used to drive new permission boundary test cases.
- **Model registry identity metadata requirements:** Require that every entry in the organization's model registry include a complete identity profile, approved permission scope, credential configuration type, delegation model, audit requirements, and current deployment approval status before the entry is marked release-eligible.

- **Post-deployment monitoring activation:** Make monitoring profile activation a mandatory step in every agent deployment, with the monitoring configuration derived from the approved identity manifest and pushed to the centralized SIEM as part of the deployment pipeline, not as a manual post-launch task.
- **Quarterly pipeline review cadence:** Schedule a quarterly review of the identity-integrated pipeline that covers gate effectiveness metrics, permission boundary test coverage, red-team finding closure rates, and deployment manifest library hygiene—retiring or re-validating any manifest configuration that has not been reviewed within twelve months.
- **Deployment classification taxonomy governance:** Define and enforce a deployment classification taxonomy based on permission scope change rather than deployment type, ensuring that any build touching credential configuration, permission manifests, or integration surface—regardless of how it is categorized—routes through the full identity review process.
- **Credential rotation tolerance testing:** Require that every agent pass simulated credential rotation testing before production deployment, confirming that the agent requests new vault-issued tokens seamlessly without service interruption when credentials are cycled.

8.6 Takeaway

Identity is not a deployment-day checkbox. It is a design-day commitment that must be woven into every stage of the AI development lifecycle with the same discipline applied to functional requirements, performance benchmarks, and data governance standards. The manager who understands this builds processes that distribute identity responsibility across the entire delivery team—from the architect who defines the identity profile to the developer who integrates the credential vault to the test engineer who validates permission boundaries to the release manager who activates the monitoring configuration. Each stage of the lifecycle has its own identity touchpoint, and each touchpoint is an opportunity to catch issues at a fraction of the cost of finding them in production. The cost curve is unambiguous, the organizational consequences of late discovery are severe, and the tools to embed identity controls into modern delivery workflows are available and mature.

The practical example of Navarro Federal Health Solutions illustrates that building an identity-integrated pipeline is a twelve-week investment, not a multi-year transformation program. The returns are immediate and measurable: fewer post-deployment corrections, faster compliance attestations, cleaner deployment manifests, and a security review process that developers support rather than circumvent because it is fast, documented, and tied to the work they are already doing. The human review gate that the Navarro team expected to create a bottleneck instead created accountability. When developers know that a security architect will read their identity manifest, they write better identity manifests. Governance embedded in the workflow shapes behavior in ways that governance applied after the fact never can.

Managers who embed identity into their pipelines do more than reduce security incidents. They build the organizational infrastructure for responsible-by-design AI governance—the operational clarity that turns pilots into trusted production capabilities and turns compliance requirements into delivery standards rather than remediation burdens. Every AI agent that enters production through an identity-integrated pipeline carries documented authorization, validated credential integration, tested permission boundaries, and active monitoring. That documentation is the foundation of trust: trust from regulators, trust from agency clients, trust from the enterprise stakeholders who must approve the next phase of AI investment. Identity governance embedded in the SDLC is not a constraint on AI ambition. It is the mechanism that makes AI ambition sustainable.

9 Operating Model for AI Identity Gov. at Scale

9.1 Scenario: The Governance Gap at Scale

Eighteen months ago, Vantara Systems had fewer than a dozen AI agents in production. A small security team knew every identity, understood every permission scope, and could trace every credential to a specific project owner. The governance model—ad hoc, relationship-driven, and largely undocumented—worked well enough that no one questioned it. Then came the growth phase. Business units across the enterprise began deploying agents independently: the customer success team launched outreach agents, operations stood up procurement and logistics automation agents, and product engineering deployed rotating cohorts of code-review agents tied to sprint cycles. By the time the CISO commissioned an enterprise-wide identity audit, the agent population had surpassed 200, spread across 11 business units, 4 cloud environments, and more than 30 distinct toolsets.

The audit findings were stark. Of the 214 agents identified, 31 had no formal registration in any identity management system—they were operating outside the governance perimeter using shared service accounts or developer API keys. Sixty-seven agents had permissions exceeding their current function. Twenty-two had not been reviewed or re-attested in over twelve months. And nine were effectively orphaned: their original project teams had been reorganized or dissolved, leaving no current owner accountable for their ongoing governance. The CISO presented this summary to the board as the single largest gap in the company's AI risk posture. The board responded with a mandate: remediate in twelve months, build a governance model that scales to five hundred agents and beyond, and ensure this problem never reaches this magnitude again.

The mandate was easy to issue and hard to execute. Vantara Systems did not have a shortage of talented people—it had a shortage of organizational infrastructure. No one owned the problem end-to-end. IAM owned identity provisioning but had no charter for agent-specific governance. Security operations had monitoring tools but no agent-specific playbooks. Compliance had audit requirements but no mechanism to enforce them on agentic systems. Product and engineering teams knew their own agents but had no visibility into the broader landscape. The CISO's challenge was not primarily technical. It was operational: how do you build a governance model—roles, processes, accountability structures, and escalation paths—that can absorb 200 agents today and remain coherent when the estate reaches 1,000?

9.2 Why It Matters

9.2.1 Ad-Hoc Identity Practices Do Not Scale

Every organization that has deployed AI agents at scale has passed through an ad-hoc phase. In that phase, individual teams make local decisions about identity provisioning, credential management, and access scoping. Those decisions are often sensible within a single project; the developer provisions an agent with the access it needs to do its job and moves on. The problem is the cumulative effect of dozens of independent decisions made without a common framework. When ten teams each operate this way, the result is ten different provisioning approaches, ten different documentation standards, ten different monitoring postures, and no one with a cross-cutting view of the aggregate. The absence of a formal operating model means identity is governed by whoever happens to make each decision, with no mechanism for consistency, escalation, or accountability.

The practical consequences of ad-hoc identity practices compound over time in ways that are difficult to reverse. Shadow agents emerge when the formal provisioning pathway is too slow, and developers use shortcuts that bypass governance controls. Credential sprawl develops as teams create credentials for testing without retiring them after the need passes. Permission drift accumulates as agents receive access for one purpose, and it is never recalibrated when their function changes. Audit and compliance gaps widen because there is no authoritative registry to audit against. And when an identity incident occurs—a credential is compromised, an agent behaves unexpectedly, a permission overstep causes a data exposure—the response is chaotic because there are no documented processes, no clear ownership, and no established communication paths. The forensic work that should take hours takes days, and the root-cause analysis that should produce a durable fix produces only a local patch.

The governance gap also has a cultural dimension that managers should not underestimate. When identity governance is inconsistent, teams that invest in responsible practices are implicitly penalized—their provisioning takes longer, their documentation requirements are more demanding, and their agents face more scrutiny. This creates a race-to-the-bottom dynamic: teams that follow the rules bear more overhead than teams that do not, and the incentive structure favors informality. A formal operating model inverts this dynamic by making governance the path of least resistance, embedding accountability into the deployment process rather than bolting it on as an afterthought. Building that model is one of the highest-value investments a manager can make in an organization's AI program.

Diagram 9.1 – Ad-Hoc vs. Governed Identity Operating Models

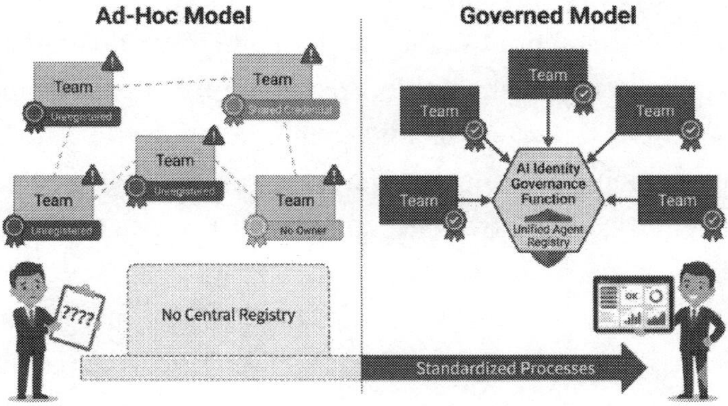

9.2.2 The Business Case for a Formal Operating Model

Governance investments are routinely challenged during budget cycles because their value is expressed in terms of risks avoided rather than revenue generated. The manager who builds a formal operating model for AI identity governance needs to articulate the business case in terms that a board, a CFO, and a business unit leader can each understand. The first driver is risk reduction. An ungoverned AI agent estate is an expanding attack surface—orphaned credentials, overprivileged agents, and undocumented identities are concrete vulnerabilities. A formal governance model systematically eliminates them by ensuring every agent has a registered identity, a scoped permission set, a documented owner, and a lifecycle process that keeps credentials fresh. The risk reduction is measurable: ungoverned agents decrease, the mean time to detect identity anomalies shortens, and incident response time falls as playbooks replace improvised reactions.

Regulatory compliance is a parallel driver of increasing urgency, as regulators begin treating AI systems as accountable entities subject to audit and documentation requirements comparable to those for other enterprise systems. Organizations that cannot demonstrate governance over their AI agents—that cannot produce a current inventory, show evidence of access reviews, or trace the lineage of a specific permission grant—face growing exposure to regulatory findings and reputational damage. A formal operating model produces the documentation artifacts regulators require: a maintained agent registry, dated access reviews, policy documentation, and evidence of escalation-path testing. Compliance is also a market-access argument: customers in regulated industries increasingly require vendors to demonstrate AI governance maturity as a condition of procurement. Organizations with a formal operating model differentiate themselves in ways that directly support revenue and partnership outcomes.

Operational efficiency and stakeholder confidence complete the business case. A governed operating model accelerates AI deployment by reducing friction around each new agent: when provisioning follows a documented process, approvals are faster, integrations are predictable, and developers know exactly what is required. Shadow agents become less attractive because the governed pathway is more reliable. The cumulative time savings across dozens of provisioning events per quarter are significant, and the reduction in incident response work—which typically consumes disproportionate amounts of senior staff time—adds further to the efficiency gains. Stakeholder confidence from the board, regulators, customers, and employees who work alongside AI agents creates conditions for broader AI adoption, greater AI investments, and a stronger organizational appetite for the kind of ambitious agentic deployments that deliver the most value.

Diagram 9.2 – Business Case for AI Identity Governance

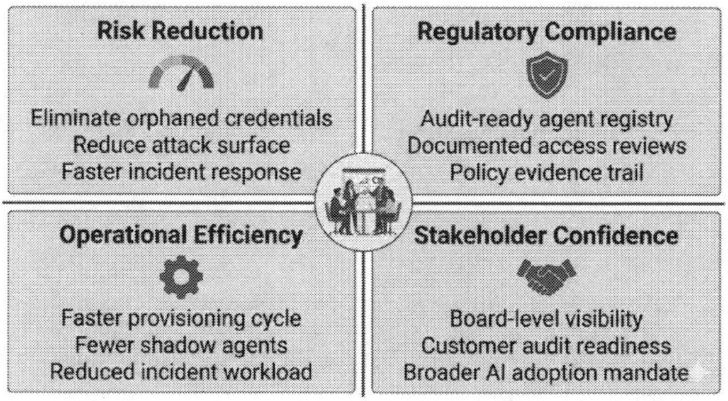

9.3 Framework: Building the Operating Model

9.3.1 Roles and Responsibilities

A governance operating model is only as durable as the accountability structures that hold it together. Defining roles and responsibilities is not an administrative formality, it is the foundational act that transforms a policy document into a functioning capability. The AI Identity Governance Lead owns the overall governance program: the agent registry, the policy framework, the RACI documentation, and the relationship between the identity governance function and executive leadership. The Governance Lead is accountable for the program's maturity trajectory—for decisions about when to invest in automation, how to design escalation paths, and how to translate governance outcomes into language that resonates with the board and the business. In most organizations, this role sits within the CISO's organization, but where IAM and security are structurally separate, it may operate as a bridge function with reporting lines into both.

Agent Identity Administrators execute day-to-day governance operations: provisioning agent identities in accordance with approved policies, maintaining the registry, executing access reviews, rotating credentials, and processing deactivation requests. In small organizations, this function can be handled by the existing IAM team with an agent-focused workstream. In larger organizations, dedicated administrators become necessary as the agent population grows beyond what a generalist team can manage without specialized tooling and process expertise. Security Operations provides the monitoring and detection layer: maintaining behavioral baselines for registered agents, analyzing anomaly alerts, executing initial triage on identity incidents, and feeding forensic findings back into the governance program. The Compliance and Audit function ensures governance practices align with regulatory requirements, manages the audit evidence trail, and prepares documentation for internal and external reviewers. Product and Engineering Liaisons serve as the translation layer between the governance function and the teams that build and deploy agents—ensuring engineering teams understand governance requirements and surfacing practical constraints that governance policies need to account for.

The interactions between these roles are as important as the roles themselves. The Governance Lead's primary interaction mode is policy and escalation—not transaction-level administration. Security Operations and the Agent Identity Administrators interact intensively during incident response, cross-referencing forensic data from monitoring with registry entries and permission records. The Compliance function interacts with all other roles on a scheduled cadence—quarterly reviews, annual audits, and event-driven reviews triggered by significant incidents or regulatory changes. Product and Engineering Liaisons are the most important relationship for long-term governance success: they are the channel through which governance requirements become engineering practice and through which engineering feedback shapes policy. Governance models that treat engineering teams as adversaries will generate shadow agents; models that treat them as stakeholders will generate buy-in.

9.3.2RACI for Agent Identity Governance

A RACI matrix operationalizes the role definitions by assigning specific accountability for each governance activity. For AI identity governance, the key activities span the full agent lifecycle: agent registration, permission scoping, credential management, monitoring, audit response, incident response, and retirement. Each activity requires clarity about who is Responsible (doing the work), who is Accountable (owning the outcome), who is consulted (providing input before decisions), and who is informed (notified of decisions after the fact). Confusion about RACI assignments is one of the most common causes of governance failure—when two functions believe the other is accountable for an activity, nothing gets done. When no function is clearly responsible, activities fall through the gaps between teams.

For agent registration, the Agent Identity Administrator is responsible, and the Governance Lead is Accountable for registry quality. Product and Engineering Liaisons are Consulted; Security Operations and Compliance are Informed. For permission scoping, Security Operations and the Agent Identity Administrator are jointly Responsible for evaluating proposed scopes against least privilege; the Governance Lead remains Accountable; Product and Engineering are consulted; Compliance is informed. For credential management, the Agent Identity Administrator is responsible, and the Governance Lead is Accountable for adherence to the rotation policy. Monitoring is Security

Operations-led, with the Governance Lead accountable for coverage across all registered agents. Audit response is Compliance-led (Responsible) with Governance Lead Accountable. Incident response is Security Operations-led, with Governance Lead Accountable; other roles are consulted or informed by severity level. Retirement is a collaborative handoff: Product and Engineering initiate the request, the Agent Identity Administrator executes it, and the Governance Lead confirms completion and updates the registry.

RACI matrices are most valuable when specific enough to drive behavior but flexible enough to accommodate organizational variation. A matrix that assigns accountability at the role level—rather than naming specific individuals—remains valid through personnel changes and reorganizations. The governance program should maintain the RACI as a living document, reviewed at least annually and updated whenever organizational changes affect the roles it describes. When the RACI is treated as a static artifact, it quickly drifts out of alignment with actual practice, and the accountability it was designed to enforce evaporates. Regular RACI reviews, combined with governance retrospectives after significant incidents, are the mechanism by which an organization keeps its accountability structure calibrated to operational reality.

Diagram 9.3 – RACI Matrix for Agent Identity Governance

Activity	Governance Lead	Identity Admin	Security Ops	Compliance	Product/ Engineering
Agent Registration	A	R	I	I	C
Permission Scoping	A	R	R	I	C
Credential Management	A	R	C	I	I
Monitoring	A	C	R	I	I
Audit Response	A	C	C	R	I
Incident Response	A	I	R	C	C
Retirement	A	R	I	I	R

9.3.3 Escalation Paths

Governance structures that lack defined escalation paths fail precisely when they are needed most—during incidents, when time pressure is highest and improvised decision-making is most likely to lead to poor outcomes. An escalation path for identity incidents is a pre-committed set of decision rules that answers three questions for each severity level: who takes action, how quickly they must act, and what conditions trigger elevation to the next tier. Defining these paths before an incident occurs is the only way to ensure they are available when needed. The alternative—escalating by judgment in the moment—means every incident response begins with a meta-conversation about responsibility, consuming time that should be spent on containment.

A well-designed escalation framework operates across four severity tiers. Informational events—anomalous but within expected variance—are handled by Security Operations monitoring without escalation: a log entry is created, and monitoring thresholds may be adjusted. Low-severity events—an agent attempting to access a resource outside its permission scope, or a credential rotation failure—trigger a Security Operations response within four hours, with the Agent Identity Administrator notified and a ticket created. If unresolved within the SLA, the issue automatically escalates to the Governance Lead. High-severity events—a credential compromise, an agent operating materially outside its behavioral baseline, or evidence of lateral movement—trigger immediate Security Operations response with parallel notification to the Governance Lead, Compliance, and the Product and Engineering Liaison. A formal incident ticket is opened, containment is authorized, and a preliminary CISO report is due within two hours. Critical events—confirmed unauthorized access, data exfiltration, or an agent acting as an attack vector—trigger the full incident response function: the CISO is notified immediately, executive escalation occurs within thirty minutes, and external counsel and regulators are notified as required. Each tier should be documented, tested through tabletop exercises, and reviewed after significant incidents.

Diagram 9.4 – Identity Incident Escalation Path

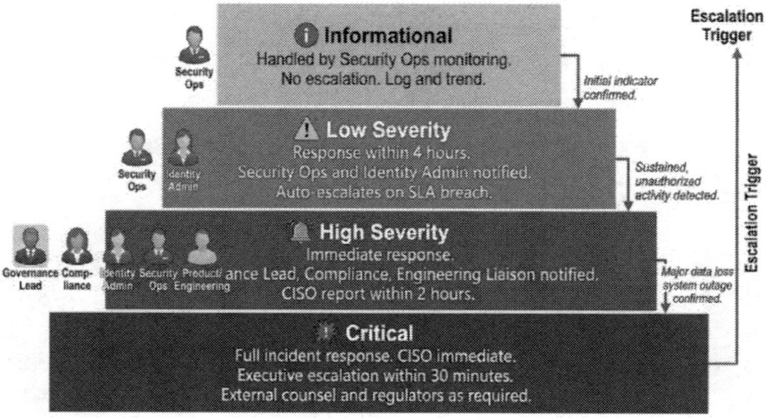

9.3.4 Cross-Functional Working Groups

No governance model survives contact with the organization without a mechanism for ongoing alignment. Cross-functional working groups are that mechanism. An AI Identity Governance Working Group brings together representatives from IAM, security operations, compliance, product management, and engineering on a monthly cadence to review governance metrics, resolve disputes about policy interpretation, share lessons from incidents and near-misses, and coordinate on upcoming deployments with governance implications. The working group is not an executive decision-making body; its function is to align and communicate. When compliance identifies a new regulatory requirement, the working group translates it into operational changes before it becomes a finding. When an engineering team finds that a governance policy creates unintended friction, the working group surfaces and resolves it before it drives teams toward shadow practices.

The discipline required to sustain a working group differs from that required to design one. Working groups fail when they lose operational focus, becoming status-reporting forums rather than problem-solving ones —or when attendance becomes inconsistent. Managers who sponsor governance working groups should set a clear agenda that reserves most of the time for discussion, track action items with owners and due dates, and maintain a visible backlog of open issues. Rotating the chair among functional representatives builds governance literacy across the organization and distributes the reputational investment in governance success more broadly. The working group should also serve as the escalation path for disagreements that cannot be resolved bilaterally. When IAM and security operations disagree on a monitoring threshold, or when product and compliance have conflicting requirements, the working group provides a structured forum for reaching a decision that all parties can implement.

9.3.5 Maturity Model for AI Identity Governance

Governance maturity models give organizations a language for assessing where they are, a framework for understanding what good looks like, and a roadmap for making structured progress. The AI Identity Governance Maturity Model has five levels. Level 1—Ad Hoc—describes an organization where agent identity governance exists, if at all, as a byproduct of individual decisions rather than a formal program. There is no authoritative agent registry, no documented governance policy, and no defined role responsible for governance outcomes. Incidents are discovered reactively, often through complaints or external reports rather than through monitoring. The primary characteristic of Level 1 is invisibility: the organization lacks a clear picture of its own agent estate and has no systematic mechanism to acquire one.

Level 2—Reactive—describes an organization that has begun responding to governance problems as they arise but has not yet built proactive structures. After a significant incident, the organization may create an ad hoc registry or assign informal ownership of identity governance. Policies may exist in draft form. Some agents have formal identities; others do not. The defining characteristic is incident-driven improvement: the organization learns from crises but lacks the infrastructure to prevent them. Level 3—Defined—marks the transition to a formal, proactive governance program. A documented policy framework exists and is enforced, an authoritative agent registry is maintained, roles and responsibilities are documented and understood, a RACI governs key lifecycle activities, escalation paths are defined and tested, and compliance reviews are scheduled and executed. Level 3 is the baseline that every organization deploying AI agents in production should hold as its minimum standard.

Level 4—Managed—describes an organization that has moved beyond documentation into measurement. At this level, key governance metrics are tracked: provisioning cycle time, mean time to detect identity anomalies, access review completion rates, percent of agents with active credential rotation, and audit finding resolution time. Governance decisions are informed by data rather than intuition, and the organization can demonstrate improvement over time. Automation has been deployed for core lifecycle activities, reducing the manual burden on practitioners and enabling the program to scale with the agent population. Level 5—Optimizing—represents the frontier of governance maturity. At this level, the program incorporates continuous improvement mechanisms: retrospectives feed policy revisions, monitoring data informs permission scoping at provisioning time, and governance outcomes influence AI program investment decisions at the executive level. Reaching Level 5 is a disposition rather than a destination: an organizational commitment to treating governance as an evolving, learning function rather than a solved problem.

Diagram 9.5 – AI Identity Governance Maturity Model

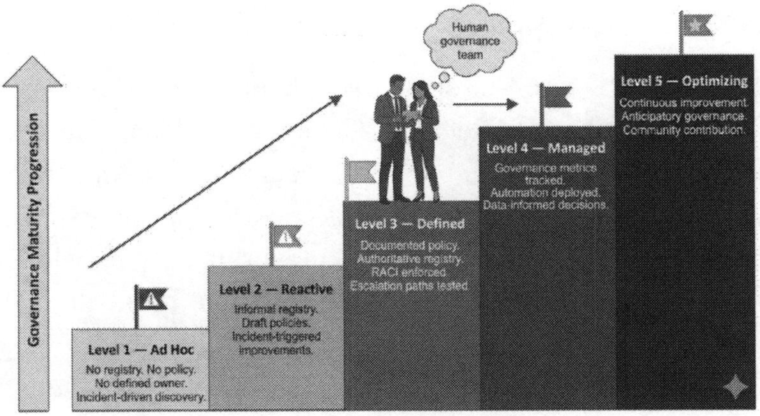

9.4 Example: 12-Mth Ident. Modernization Plan

Returning to Vantara Systems: after the board mandate, the CISO assembled a cross-functional team—the IAM lead, a senior security architect, the compliance director, and two engineering leads from the largest-agent business units—to design the remediation roadmap. Their first decision was to resist the temptation to begin with tooling, which often produces expensive implementations that automate poorly understood processes. The team committed the first three months entirely to discovery and assessment. Deliverables for this phase included an authoritative agent inventory covering every agent across all business units and cloud environments, a current-state map of existing governance capabilities, a gap analysis that translates the map into prioritized remediation actions, and a board-approved business case. The manager's role in this phase is to clear the path: ensuring that business unit leaders cooperate with the inventory and engineering teams to grant the audit team the necessary access, and that findings are presented honestly rather than minimized.

Months four through six built the governance foundation. The AI Identity Governance Lead role was created and filled, agent identity administrator responsibilities were formally assigned within the existing IAM team, and the cross-functional working group held its inaugural session. RACI documentation was drafted, circulated for feedback, and published as a governing document. A policy framework was developed covering agent registration requirements, permission-scoping standards, credential-rotation schedules, and acceptable-use criteria for shared-service accounts. Tooling evaluation, understanding what was available on the market, what existing tools could be extended, and where net-new investment was required—ran in parallel with policy development. The deliberate sequencing mattered: by the time tools were evaluated, the team had a clear picture of what processes those tools needed to support, preventing the evaluation from becoming a vendor-led requirements exercise. The measurable outcome of this phase was a governance charter specifying the program's scope, authority, and operating model, which every functional lead had reviewed and formally committed to.

Months seven through nine were implementation months. The agent registry was deployed using existing identity governance tooling extended with a purpose-built agent tracking layer, and every agent from the Month 1–3 inventory was migrated into the registry with a confirmed owner, documented permission scope, and active lifecycle status. Lifecycle automation was implemented for the highest-volume activities: automated credential rotation, automated deactivation triggered by project-closure events in the organization's project management system, and automated alerts for agents who did not re-attest within the policy-defined review window. The monitoring infrastructure was extended to include agent-specific behavioral baselines, establishing the detection foundation on which the escalation path would depend. Training was delivered to all product and engineering teams with active agent deployments—not as a compliance checkbox but as a structured engagement designed to make governance literacy part of engineering culture. The manager's role in this phase is quality control: ensuring that implementation delivers what the policy framework requires, not just what the tooling makes easy.

Months ten through twelve operationalized the governance program. Audit workflows were activated, with the first formal access review producing a compliance report demonstrating the program's ability to generate the evidence artifacts that regulators and internal audit required. The escalation path was tested through a structured tabletop exercise simulating a high-severity identity incident, with all role-holders participating and results documented for post-exercise improvement. A formal maturity assessment produced a score of Level 3 (Defined)—a significant advance from the Level 1 baseline at the start of the year. The final deliverable was a roadmap for advancing to Level 4: a prioritized set of investments in governance metrics, automation expansion, and continuous improvement mechanisms, presented to the board alongside the maturity assessment as evidence that the remediation mandate had been fulfilled and the program had the infrastructure to sustain its progress.

What made this roadmap successful was not technical sophistication; it did not require novel tooling or unusual expertise. What made it successful was disciplined sequencing: discovery before foundation, foundation before implementation, implementation before operationalization. Each phase built on the one before it and produced artifacts that the next phase depended on. Organizations that skip discovery and go directly to tool procurement often automate processes that are not yet understood. Organizations that define policies without establishing roles find that policies sit unread because no one owns their enforcement. The manager's contribution at each phase is organizational discipline: ensuring that the sequence is respected, phase deliverables are completed before the next phase begins, and cross-functional stakeholders cooperate as required. A twelve-month governance roadmap is, at its core, a change management program. The technical work is necessary but not sufficient. Organizational work is where governance programs succeed or fail.

Diagram 9.6 – 12-Month Identity Modernization Roadmap

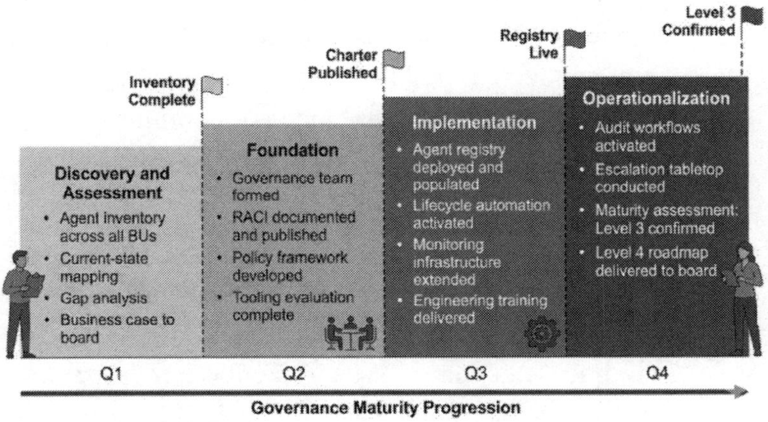

9.5 Manager's Checklist

- **Governance team charter documented and approved:** Formal charter defines the governance program's scope, authority, reporting line, and decision rights. The charter has been reviewed by the CISO, legal, and relevant business unit leaders and is published as a governing document accessible to all stakeholders.
- **RACI matrix completed and distributed:** A RACI matrix covering all key agent identity governance activities—registration, permission scoping, credential management, monitoring, audit response, incident response, and retirement—has been documented, reviewed by all role-holders, and

published. A review date is set for the next calendar year.

- **Escalation path defined, documented, and tested:** Escalation paths for identity incidents across all four severity tiers are documented with response time SLAs, escalation triggers, and role-specific responsibilities. At least one tabletop exercise has been conducted to validate that the paths function as designed under simulated incident conditions.
- **Cross-functional working group established and on schedule:** A monthly AI Identity Governance Working Group with representation from IAM, security operations, compliance, product, and engineering has held its inaugural session. A standing agenda format, action-item tracking process, and rotating chair schedule have been established.
- **Maturity assessment baseline completed:** The organization has conducted a formal assessment against the five-level AI Identity Governance Maturity Model and documented a current-state score. A target maturity level for the next twelve months is established and included in the governance program roadmap.
- **Agent inventory complete and populated:** An authoritative inventory of all AI agents in the organization's environment has been completed, covering all business units, cloud environments, and toolsets. Every identified agent has been entered into the agent registry with a confirmed owner,

documented permission scope, and active lifecycle status.

- **Policy documentation published and enforced:** A policy framework covering agent registration requirements, permission scoping standards, credential rotation schedules, shared-service-account usage restrictions, and acceptable-use criteria is published and actively enforced. Policy exceptions are documented and approved through a defined exception management process.
- **Tooling requirements defined and procurement underway:** The governance program has documented its tooling requirements based on the policy framework and operational processes it needs to support. A tooling evaluation has been completed, and a procurement or implementation plan is in progress, with an expected go-live date tied to the implementation phase of the roadmap.
- **Training plan developed and delivered to engineering teams:** A structured training program for product and engineering teams covers agent registration requirements, the provisioning process, credential handling standards, and escalation procedures. Training completion is tracked, and refresher training is scheduled at least annually or when significant policy changes occur.
- **12-month roadmap published with milestones and owners:** A documented twelve-month governance roadmap organized into four quarterly phases—Discovery, Foundation, Implementation, Operationalization—is published with specific

deliverables, milestone dates, and named owners for each deliverable. The roadmap has been reviewed and approved by the CISO and relevant business unit sponsors.

- **Quarterly governance review cadence established:** A quarterly governance review process is scheduled and confirmed, bringing together the Governance Lead, CISO, compliance director, and working group chairs to review program metrics, assess roadmap progress, and make resource and priority decisions for the coming quarter.
- **Audit evidence trail verified and audit-ready:** The governance program's documentation artifacts—agent registry, access review records, policy versions, exception logs, incident records, and training completion records—are organized, current, and confirmed by internal audit to be sufficient for regulatory examination. A formal evidence review has been scheduled at least annually.

9.6 Takeaway

A scalable operating model for AI identity governance is not a luxury that organizations can defer until they reach some threshold of size or sophistication. It is the organizational infrastructure that determines whether AI adoption is sustainable or chaotic—whether the agents an organization deploys are a governed, auditable, risk-managed capability or an expanding patchwork of ad-hoc decisions that accumulates liability with every new deployment. The Vantara Systems scenario is not unusual; it is, in various forms, the situation most organizations will face if identity practices scale faster than governance structures. Managers who understand the stakes do not wait for a board mandate. They begin the governance work before the crisis, because the cost of building governance proactively is a fraction of the cost of remediating a governance failure at scale.

The operating model described in this chapter—roles and responsibilities, a RACI matrix, escalation paths, cross-functional working groups, and a maturity framework—is not a bureaucratic overlay on top of technical work. It is the mechanism by which technical governance decisions become organizational capability. A RACI matrix is only as valuable as the accountability culture that enforces it. An escalation path is only as reliable as the rehearsal that validates it. A working group is only as effective as the discipline that keeps it focused on problems rather than presentations. The manager's contribution to AI identity governance is organizational leadership: creating conditions in which the right people do the right work with clear accountability, sufficient information, and a shared understanding of what mission-aligned governance looks like in practice.

Organizations that invest in this infrastructure position themselves to govern AI identity at any scale—to absorb growth in their agent estates without corresponding growth in governance risk, and to demonstrate to regulators, customers, and their own boards that their AI programs are responsible by design rather than compliant by accident. The twelve-month roadmap is a starting point, not a destination. Governance maturity is a continuous journey, and the organizations that understand this—that treat governance as a capability to be developed rather than a problem to be solved once—will sustain the trust required to deploy AI at the scale and ambition that the technology makes possible. That ambition, grounded in operational clarity and held accountable by a mature governance model, is what transforms an AI program from a series of experiments into a durable, mission-aligned enterprise capability.

10 Conclusion

AI identity governance is no longer a niche technical concern—it is the backbone of safe, scalable, mission-aligned AI. Across this book, you've seen how quickly identity failures can derail even the most promising initiatives. As one early chapter warns, *"The technology had worked. Identity had failed, and identity had taken the whole project with it."* That pattern repeats across industries: the agent behaves exactly within the permissions it was given, and the organization discovers—too late—that those permissions were never designed for autonomous, high-velocity systems.

The chapters you've worked through form a complete operating system to prevent those failures. You learned why traditional IAM assumptions collapse under agentic workloads, how to classify identities correctly, and how to architect trust boundaries, token flows, and delegated authorization to enforce governance at machine speed. You saw why lifecycle discipline matters—because without it, "zombie agents" and shadow identities accumulate silently until they become attack surfaces. You explored the six identity risk categories unique to AI and multi-agent systems, and why even small gaps can cascade into compliance events that halt entire programs. And you built toward a mature operating model—roles, RACI, escalation paths, working groups, and a roadmap that scales from dozens to hundreds of agents without losing control.

If there is one unifying lesson, it is this: identity is the control surface of AI. Everything an agent does—every query, every write, every cross-system hop—is an identity event. When identity is intentionally designed, continuously monitored, and governed structurally, AI becomes a force multiplier. When identity is improvised, inherited, or treated as an afterthought, AI becomes a generator of liabilities. The difference is not technical sophistication; it is managerial discipline.

Your role as a manager is to make that discipline real. That means insisting on scoped, ephemeral credentials; requiring identity profiles at design time; embedding identity gates into CI/CD; demanding audit-ready logs; and ensuring every agent has an owner, a lifecycle, and a retirement plan. It means building a culture where teams understand that identity is not a security tax—it is the enabler that enables AI to operate safely in high-stakes environments.

AI will only grow more autonomous, more interconnected, and more central to mission-critical workflows. Organizations that build identity governance now will move faster, avoid costly remediation cycles, and earn the trust of regulators, executives, and the public. Those who delay will find themselves reacting to incidents rather than shaping their AI future.

You now have the frameworks, language, and operational patterns to lead from the front. The next step is simple: choose one governance improvement, implement it well, and build from there. Identity is the foundation. Build it first, strengthen it continuously, and your AI program will have the stability and credibility it needs to scale with confidence.

11 Works Referenced

1. European Commission. *Regulation (EU) 2024/— on Artificial Intelligence (AI Act)*. Official Journal of the European Union, 2024. (CH9)

2. Federal Trade Commission. *Protecting Americans from Deceptive or Unfair AI Practices: Staff Report and Recommendations.* U.S. Federal Trade Commission, 2024. (CH6)

3. National Institute of Standards and Technology. *Artificial Intelligence Risk Management Framework (AI RMF 1.0).* NIST Special Publication NIST.AI.100-1, Jan. 2023. (CH7)

4. National Institute of Standards and Technology. *NIST AI RMF: Generative Artificial Intelligence Profile (NIST-AI-600-1).* NIST, July 26, 2024. (CH5)

5. Office of Management and Budget. *Memorandum on Safe, Secure, and Trustworthy Development and Use of Artificial Intelligence.* Executive Branch Guidance, Oct. 2023. (CH8)

6. ISO/IEC. *ISO/IEC 42001:2023 — Artificial Intelligence Management Systems — Requirements.* International Organization for Standardization / International Electrotechnical Commission, 2023. (CH9)

7. European Union Agency for Cybersecurity (ENISA). *Threat Landscape for AI Systems: Risks, Vulnerabilities, and Mitigations.* ENISA Report, 2024. (CH6)

8. U.S. Department of Commerce. *AI Accountability and Governance: Implementation Guidance for Federal Agencies.* Departmental Guidance Paper, 2024. (CH9)

9. Carnegie Mellon University Software Engineering Institute. *Workload Identity and Ephemeral Credentials for Cloud-Native Systems.* SEI Technical Report, 2023. (CH3)

10. IEEE Standards Association. *IEEE P7002 / P7003 Series: Data Privacy and Algorithmic Bias Guidance (Selected Reports).* IEEE, 2023–2025. (CH6)

11. ACM Conference on Computer and Communications Security (CCS). "Identity and Access Control Challenges for Agentic Systems." Proceedings of ACM CCS, 2024. (CH3)

12. US Department of Health and Human Services. *Guidance on AI and Health Data Privacy under HIPAA.* HHS Office for Civil Rights, 2024. (CH6)

13. Microsoft Research. *Delegated Authorization Patterns for Autonomous Agents.* Microsoft Research Technical Report, 2024. (CH4)

14. Google Cloud. *Workload Identity Federation: Design Patterns and Security Considerations.* Google Cloud Whitepaper, 2023. (CH3)

15. Amazon Web Services. *Best Practices for Short-Lived Credentials and Token Management in High-Throughput Systems.* AWS Security Whitepaper, 2023. (CH5)

16. Center for Security and Emerging Technology (CSET). *AI Agents and the New Identity Attack Surface.* CSET Research Brief, 2024. (CH6)

17. Brookings Institution. *Governing AI: Policy Tools for Identity, Accountability, and Auditability.* Brookings Report, 2023. (CH9)

18. Journal of Cybersecurity (Oxford). "Cross-System Auditability for Autonomous Agents: Methods and Case Studies." *Journal of Cybersecurity*, vol. 9, no. 1, 2024, pp. 45–62. (CH7)

19. RAND Corporation. *Operationalizing Explainability and Audit Trails for Agentic AI in Government Systems.* RAND Research Report, 2025. (CH7)

20. SANS Institute. *Identity-Aware Monitoring and Incident Response for AI-Driven Workloads.* SANS Whitepaper, 2024. (CH7)

21. Oxford Internet Institute. *Identity Drift and Shadow Agents: Empirical Findings and Governance Implications.* OII Working Paper, 2025. (CH5)

Made in the USA
Monee, IL
10 November 2026

86ca8c6f-2356-4ced-90af-f54284944426R01